GRAND PHASES ON THE
SUN

The case for a mechanism responsible for extended solar minima and maxima

STEVEN HAYWOOD YASKELL

Order this book online at www.trafford.com
or email orders@trafford.com

Most Trafford titles are also available at major online book retailers.

Printed in the United States of America.

ISBN: 978-1-4669-6301-6 (sc)
ISBN: 978-1-4669-6300-9 (e)

Trafford rev. 12/21/2012

 www.trafford.com

North America & international
toll-free: 1 888 232 4444 (USA & Canada)
fax: 812 355 4082

For my father, John Yaskell, Sr.

Contents

Acknowledgements...ix

Only eight light minutes away..xiii

1. Comprehending, and contemplating, deep time...1
2. What is a grand solar phase versus a "regular" solar phase? 10
3. Grand solar phases in possible civilization-altering contexts 31
4. Widening perspectives of our Sun in space: what does the Sun and other
   star phenomena produce in the Sun-earth climate connection—and what
   could it do to Earth?...61
5. "Centers of activity" on the Sun: a linear view of the nonlinear as an
   introduction to helioseismology 83
6. Total solar energy and particles, climate, Earth's orbital considerations,
   and the solar dynamo in the de Jager-Duhau synthesis..........................102
7. What a grand solar phase mechanism might reveal in the short term.......127
8. A summary, some observations, and some closing thoughts 151

Acknowledgements

This book comes on the eve of the 2013 solar maximum of Solar Cycle 24. It outlines in a speculative and probing manner the approaches to how the Sun may be going into either an extreme grand phase (higher activity; a grander phase than it presumably already has been in or just as much so)—or even into a polar opposite extreme: a grand "episode" or grand minimum. That is, the coming solar maximum might be so weak it would almost be like a minimum. A middle area between both extremes might turn out to be the case. Areas of the most advanced research in several disciplines have been investigated, riffled through, described, argued over and then laid out for circumspection like a hundred pieces of a strange tool that has not been accompanied by any documentation on how it is assembled or how it works. At the same time, some ground not always thoroughly considered in these matters is covered, to include its history. For some it will be an obvious reiteration of old themes and basic science yet to others, it will be revelatory (as much of it was for me). In any case, to all members of Robert Boyle's Invisible University in the republic of letters, now is a golden time to be circumspect.

Will 2013-14 be an even larger extension of the putative Modern Maximum (c. 1924-2009) or turn out to be an incredible dud? The object of the exercise has not been merely to satisfy the Muse of Curiosity, *per se*. It has not been for large sums of money, a grant extension, or for any group or actor, governmental or otherwise. Journalistic and pundit enthusiasm aside, fomenting this skirmish in the so-called science wars for various reasons (politics, positions, larger grants, fixed beliefs etc.) had one positive unintended side effect. It opened up a chance to peel back the layers of our Sun to see if it betrayed a deeper regularity than heretofore known. What can we learn from something like this? What can it give us in terms of practical versus purely esoteric benefit? Would such knowledge help gain us a measure of security in view of some of the famous tantrums for the weak or extra-powerful our Sun can from time to time throw? Would the garnered knowledge help us some day to get past or at least revolve safely around it?

Three people in particular shaped this project. Approaching Cornelis ("Kees") de Jager, I think the last living solar scientist to have all major medals awarded in the field to him, was not so daunting as it was promising. His volunteer research of late intrigued me. This long-retired laborer in the field of solar research's record is awe inspiring to say the least. He is perhaps the world's foremost authority on solar flares. He has the Gold Medal from the Royal Astronomical Society (UK); the George Ellery Hale Prize (AAS, United States) for solar research; the Jules Janssen Medal (France) for solar research and the Karl Schwarzschild Medal (for astrophysics) (Germany). He has the Gagarin Medal and Ziolkowski Medal (space research, former USSR). The list of awards is a page or so long. His former titles range from being president of the International Council of Scientific Unions (ICSU . . . now the International Council for Science); a past general secretary of the International Astronomical Union (IAU) and a president of COSPAR (International organization for co-operation in space research). Perhaps he achieved all this since he received his PhD in 1952—a long way before grade inflation overtook the academy. He is part of functionalist science, common from the 1920s to 1950s but now sadly out of vogue in these times of increasing formalism and clerical obsession in both institution and school. Take heed, for we may not hear of the likes of one such as him for a very long time. Formally retired from his field and from some of the institutions he even founded, he has since 2003 been a volunteer researcher at the Royal Netherlands Institute for Sea Research. He has concentrated mainly on the study of the Sun-climate relationship there. He brings to bear upon musing on this relationship almost thirty years of expertise in the structure and the dynamism of the Sun's so-called atmosphere, *alone*. In talking and working with him for this brief time, probably the last such hand-off from one so experienced and honored to one so naïve and not honored in such an area, I personally hope for much wheat being separated from chaff in the mind fields we both so carefully stepped around and inside of. Of any liberties or license unintentionally introduced here he is innocent. He brought solar theory to bear in the Sun-earth relationship regarding mechanisms for extended grand solar phases. Further considerations are the work of myself and the other collaborators, of whom I shall address forthwith.

At some point around 6,000 miles above us what is the Sun becomes Earth and Earth, Sun. Kees needed an upper atmospheric chemist and dynamics specialist of some boldness, dedication to honest research, and ingenuity to round it out and found it in the Argentinean, Silvia Duhau. She was a former fellow at the Laboratory for Upper Atmospheric Physics of the NASA Goddard Space Flight Center. Silvia founded the laboratory of geophysics in the physics department of Buenos Aires University, much like Kees had established an institute in the Netherlands. She has pioneered studies in geomagnetism and geophysical prospecting and also has to her credit several awards. Then, needing an environmental scientist, Bas van Geel at the University of Amsterdam came forth as a Quaternary paleoecologist/paleoclimatologist. Far from the sun, he is a sublunary denizen of Earth's fens, bogs, and Holocene glacial deposits in search of proxy isotopic evidence in these "archives" or reservoirs. His work with W.G. Mook brought in precision wiggle-matching to date organic deposits such as climate-sensitive peat with more accuracy than previously seen, bringing into sharper focus the overlay of changes in C14 atmospheric concentration from changing solar activity and such deposits. An enthusiastic supporter of this project as much as Cornelis, I shake his hand (by proxy, of course). I shake by distance Kees' and Silvia's as well and thank all for technically editing and advising over what is an attempted written amalgam of their contributions to science. To whatever failure this leads in transmission (and it always does somewhere) I as the writer take full responsibility. In my defense, I found many and various reasons and justifications as to why, when approached, highly qualified teachers and practicing scientists could not offer technical editing services. I was taken aback to learn that some felt unqualified to check the physics I have tried/ have recorded, in these pages. Sensing fear I backed off from others I had in mind for the task. Fools tread where angels fear to go (or at least those without a professional career to jeopardize). A fourth contributor is another impendent researcher. This would be William Neil (Bill) Howell of

Ottawa, Canada (not far from my Alma Mater at Carleton). I heartily thank him and his father for their personal and private intellectual (and financial) contributions to this work.

This is not an easy book to read. However the scientific papers from which much of it was derived are in many cases beyond the lay reader and had to be translated, as it were. The connections between those papers in this attempted translation are even more hazardous—which explains in part the reluctance of some potential editors hearing my appeal. The book is mainly information: entertainment is a secondary concern. With this in mind, I most fervently hope the implications for climate change for the cooler (and over a longer period as projected) in this work prove entirely untrue and that a continuation of hemispheric warming as promised from some quarters be truly the case. Whatever prevails, I pray that the putative mechanism behind grand solar phases de Jager and Duhau have delineated be further unraveled in the fruitful give and take of reasoned and good-natured scientific falsification for any good which it might obtain. Falsification is the most progressive—as well as potentially most utilitarian and ultimately most humanitarian—aspect of all scientific investigation.

Steven Haywood Yaskell
November 16th 2012
Belmont, Vermont, USA

Only eight light minutes away

The Sun: a type G 2 V star, main sequence.

This means it is still "young" in that it has enough (it is thought) nuclear fuel to bathe us in light and life-giving warmth into time farther than we or our distant descendants can imagine. Like stars are wont to do the Sun will most likely implode and vanish one day. That's what happens to all we imperfect humans perceive in time. But the Sun also has maximum and minimum phases revealing that it varies in luminosity and so, a certain strength. It is variable.

Then there's the distance. Mentioning distance in astronomy, besides habit and mathematical fetish, sometimes has extremely useful application.[1] As regards the Sun, it is 149,000,000 kilometers or 9,2584,307.643 miles away from us. This latter is an untidy figure close enough to the tidier 93 million miles that gives rise to that even tidier little measure, one (1) Astronomical Unit (A.U.). The A.U. is used with detailed precision by professional astronomers everywhere. It takes a bit more than 63,000 A.U.s to make up a single light year. One light year [2] is almost six trillion miles. So as regards stellar distance and closeness, the Sun is *close*. The Sun, then, is only eight light minutes away from us.[3] Such is the indifference with which we toss out numbers and facts about *Sol,* A.K.A the Sun, the closest star to us unless there is a hidden one somewhere that would cause our Solar System to be known as a binary star system. [4]

There are the numbers in astronomy, so vital, and now, so conflicting and easy to obtain and compare and almost always too deep to comprehend except for a very few. The strangest thing we have to accept with all these figures is, when we think of such matters in a detached-from-physics, very personal manner, we have to forget the exact numerical details while, paradoxically, holding desperately on to them in perpetual concatenation, if they are of immediate and direct use. The numbers, howsoever many conflicting, working themselves back and forth in time, like the words in the theories they are attached to, give us a clearer picture as we gain in knowledge, if in a different light. But they also blur.

---

[1] Often it is arcane, theoretical and even bizarre except to a fraction of specialized researchers.

[2] Light years are used alternatively with the stellar distance measure called the parsec. One parsec is almost 20 trillion miles, or about 3 ¼ light years.

[3] In seconds, it is 149.6 / 0.3 = 499 light-seconds. And this equals 8 light minutes and 19 light seconds.

[4] There is speculation that our sun is part of a binary system. That is, there is another star near us. This was promoted by Alfred de Grazia and Earl Milton in *Solaria Binaris, Origin and History of the Solar System* (Mertron: 1984).

This applies to the multifarious concepts in astronomy as well. For the persons to whom this book is mostly aimed, something like knowing what the Sun's "G class" involves is a desperate gulf between those who know and who do not know ("V" is for variable). We step with trepidation over the chasm of "what does that mean?" let alone matter. That "G" is a kind of refinement or correction, as in the now dyed-in-wool universal stellar spectral analysis and classification of O, B, A, F, G, K, M, from the original A, B, C, D, E, F, G, (etc.)—which was figured out later to be wrong if a good first try.[5] In the process of making it have more technical sense the order got all mixed up. But these key letters in spectral analysis—literally how much particle excitations are prominent in a star's light, and so identifying what precisely some stars are elementally made of versus others—was a difficult enough complexity of human thought to begin with. All this is a strong sign that humans proceed in a very trial-and-error way with complex things they learn. This whether it was our distant ancestors first chipping a stone just right to make a tool that wouldn't fail in a hunt, to astrophysics. It is the way we have progressed, from crushers of lion's heads to crushers of atoms—something "He" (Sol) does, too. [6]

The Sun, lastly in this vein, is also what is called a "main sequence" star. As a "G" that means it is right in the middle of a chart American mathematician-astronomer (H.N. Russell[7]) and collaborator Danish astronomer (E. Hertzsprung) built tentatively off the pioneering distance-to-stars-from-us work of Henrietta Swann Leavitt and other women researchers at Harvard College Observatory around the year 1911. It was to be called the Hertzsprung-Russell diagram. Painstaking study of photographic plates made in South America got pored over by Leavitt until she saw a regularity in variable stars in the constellation Cepheus. These measuring stars, or "standard candles," came to be called Cepheid variables and were the first mathematically-observable sign that stars—variable or less so—betrayed some regularity or predictability. This was something that had become less mysterious by the time proper stellar motion had been statistically analysed and measured. Apparently there was some regularity even to stellar energy as well as motion. Capricious Nature in this regard could not be that capricious after all.

The Hertzsprung-Russell diagram has much to do with our particular sun is its nuclear strength at this time and how it gives off its light in the burn of the elements it consists of. As such our sun is one of a class of Sun-like stars as they now say, and these stars, many extremely far away, are studied with the full understanding that, as they behave, so does our sun possibly behave. By proxy then, these Sun-like stars help us to learn things about Sol. Yet learning by proxy as

---

[5]   William Huggins (England) Henry Draper (USA: his quartz prisms isolating Vega's "pure spectrum" photographically) Angelo Secchi (Italy) Lewis Rutherfurd, Antonia Maury, and Annie J. Cannon. When difficulty arose in accounting for elements other than Hydrogen and Helium, it was Cannon who recognized the subtle gradations in the elements in certain letter classes that gave the stellar spectra an easier to account for (and *technically* understandable) order. A good discussion of this is found in Pannekoek, A., *A History of Astronomy* (Allen and Unwin:1961) pp 451-460.

[6]   In the old literature on the Sun it was a "He."

[7]   Henry Norris Russell united luminosity with stellar classification.

you will see in this book brings its own hazards. Additionally, such added knowledge can fog up as much as to sharpen, focus.

Henrietta Swann Leavitt (1868-1921) discoverer of logarithmic period-luminous variability in stars, thereby securing the first step in accurately measuring deep stellar distance.[8]

The time has arrived to give the tossed-out distance, "eight light minutes away" more seriousness than it probably deserves. And though it has been argued that all suns are now variable to a certain extent, including ours (possibly one of the pulsating variety since Sol expands and contracts and oscillates at different speeds at some times more than others) it is time to give this fact more attention as well. Across these pages you will meet with the sense and argument of what our star is as a variable entity. This in order, hopefully, to built a structure toward obtaining a useful idea as to how it affects us in its normal and more importantly, less normal, "moods."

As the English pastor-poet John Donne once wrote about the then-cutting edge work of his Italian contemporary, Galileo, to effect: "he brought the stars closer to us so that they could speak more clearly to us about themselves." We humanize the object (the Sun) to make it familiar while denying its humanity since it has none. It is object of spiritual focus and mental dissection. So have we pulled stars including our own down a little nearer since then, the Sun and exploding stars—supernovae—toward the Earth. And lo and behold, stepping back and taking a look, we seem to have gotten even smaller in the great scheme of things. This perspective should no longer produce fear. Certainly not in a space age with an educated population far outstripping that which existed but half a century ago. It does not defy comprehension by an open, honest, and enlightened mind. Yet I understand that the glut of information available on the Sun sometimes strips bare the open attempts to describe it in an operational totality with words and numbers, as the demonstration with spectral analysis and H-R diagrams above just illustrated. Experts find and toss out: the challenge to comprehend in a holistic manner is left to the rest. Males pioneered ideas, insights, technology and accumulated much data on stellar

---

[8]   There is some evidence that Leavitt was to be awarded the Nobel Prize for physics in 1912 on recommendations by the Swedish Academy due to this discovery.

objects. Yet both the spectral class analysis tool and the logarithmic law of stellar distance in their first practical applications were hit upon by females who pored over their gleanings. I wish to stress this fact and underline it.

The wideness of the Solar System and what is beyond is clear, real, and fearsome. On the positive and upbeat modern side, space observatories like Hubble have given us not only knowledge but even a certain joy in the immensity of it all that urges us onward yet to *dare*. The crystal spherical perfection-view beyond our Solar System has been on the wane since all roads led to Rome, though this view had never been defeated for all time. Clear and daunting ancient thinkers like Eudoxus, Ptolemy and others had long ago given us an ever-widening appreciation of the vastness of space. Kepler, Gilbert, Descartes, and Newton put it all into motion on a magnetic carpet. Without the Medieval-period Muslims Al Tusi and Ibn Al Rushid the measurements and motions would possibly still be mysterious in certain aspects.[9] We become aware of our connection to near and deeper space, and must be made cognizant of how this connection effects us in at least a preliminary manner regarding the Sun—even if error will daunt each step in many crucial and even heartbreaking ways. The paradox of seeing more and deeper into the universe giving us personal insecurities of lessened importance or size is clear. [10] That forces from us and to the Sun—the electromagnetic/geomagnetic force—can effect power grids, supply, radio waves and so on is known, documented, and not fully understood. It can and will irradiate the space voyager in ways still anomalous, and the benefits here are not totally accepted let alone known well. That the Sun conspires to affect Earth's weather and oceans—its climate—is very unsettling, controversial, and yet is being pieced together. Many paths for further research are open here, and they should be thoroughly and honestly pursued for our well being in the future.

The Sun still blinding us metaphysically in this late age embarrasses some. To many it is an omnipotent all-knowing alternately unknowing, force. For others the Sun is a challenge that must be taken on and a force to understand well, and to even bypass one day.

---

[9]   See North, J., "Western Islam and Christian Spain," in *The Norton History of Astronomy and Cosmology*, (Norton:1995)

[10]   The beguiling experience of the widening of the human mind in such philosophical terms (perhaps even evolutionary ones) is lucidly covered in the pitifully obscure first book of an intended series, *The Fabric of The Heavens: The Development of Astronomy and Dynamics* by Stephen Toulmin and June Goodfield (Harper: 1961).

Improvements in photographing stellar spectra allowed for more refined study of it. (Left) Annie Jump Cannon (1863-1941) delineator of the first successful spectral class analysis table from the data. (Right) Antonia C. Maury, whose work on the same (at Harvard) regarding setting straight confusion with Helium and other problems enhanced Cannon—and future science—in her work.

It is, after all, just a few (light) minutes away.

◊

We look at the Sun for clues as to how it may or may not work. Take the following pithy technical description from some time ago:

I liken the sunspots to clouds or smokes. Surely if anyone wished to imitate them by means of earthly materials, no better model could be found than to put some drops of incombustible bitumen on a red hot iron plate. From the black spot thus impressed on the iron, there will arise a black smoke that will disperse in strange and changing shapes. [11]

We absorb this concrete and straightforward description of Galileo on what sunspots "are" from 1612 and note that, so far as scientific observation and proof was concerned, the description was as technically exact as nearly any for 300 years. Note well that Galileo also burdens his readers with attempts at experiment to widen the observation's understanding by demonstration with "earthly materials," appropriately enough such as small bits of coal daubed onto a hot iron plate to obtain the expected result. Coal was to become a major power source much later than when this description was first made (and it still is) which makes Galileo's insight here all the more fascinating. He displays both scientific imagination and, I emphasize, intellectual courage.

---

[11]   Galileo, "Third Letter on Sunspots, from Galileo Galilei to Mark Welser, In which Venus, the Moon and Medicean Planets are also dealt with, and new appearance of Saturn are revealed," (December 1, 1612) in *Discoveries and Opinions of Galileo*, Trans. by Stillman Drake (Doubleday:1957) p. 140.

The real "death" of Galileo's observation on sunspots and what they are, and actually do, begins with the Scottish engineer / scientist William Thomson (later Lord Kelvin), who will appear frequently in this book in the many different operating poses of the working scientist. I ascribe it to him since, in what are perhaps famously his own words, to effect: "when you can measure something, then you know more about it." But as you will see, "this death" toward the end of the 19th Century is only the beginning of Galileo being falsified. Since Kelvin (as he came to be called) was a science chef who whipped up the only well-cooked meal on this investigation of Sol that was then possible to cook, all the *sous chefs* who supplied the ingredients and served the side dishes are largely forgotten. Since practical science often needs not the memory, it is just as well. But since practical science generally goes nowhere without the hundreds of lesser scientists, savants, metaphysicians and assorted dreamers who supply the master chefs with the vital ingredients for ultimate realization of a great meal, these forgotten souls—usually with their moving little stories of forbearance in tow—bear passing notation.

First there was a need to see stars as matter in the proper perspective. The earliest Greeks aside, any idea for "island universes" as they are termed dates to the mid to late 18th Century and the idea of matter coalescing into groups or clumps in space. Visually these metaphysics were derived from the telescopic investigations of for example Jean-Philippe Loys de Chéseaux of Switzerland and Charles Messier of France and significantly the German-English crossover Wilhelm (later William) Herschel. Large-aperture telescope maker Herschel is especially of note in providing descriptive and measured data along with his sister, Caroline on hitherto unseen stars and star clumps and matter that was nebulous but inexplicable.

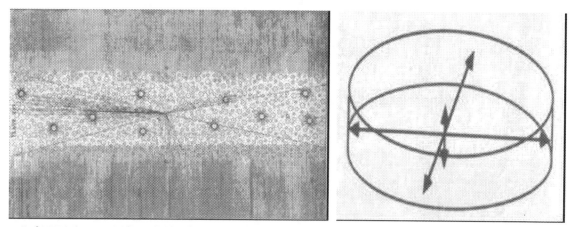

(Left) Wright's optical qualities of an island universe where stars are visible in a 360 degree view, or, all around, from a central point. (Right) Diagram of the idea of humans at the locus (middle) of a starry plane or disc as later interpreted by Kant (after Crowe [12])

[12]   Crowe, M. J., *Modern Theories of the Universe, from Herschel to Hubble* (Dover:1994) pp 30-34

Clockmaker-become-cosmologist/cosmogonist Thomas Wright proposed what could be termed a grand theory of this in his *Original Theory or New Hypothesis of the Universe* (1750). Like Kepler, who was part astrologer (and his Three Laws of Planetary Motion were originally speculations on improving astrological principles) the fantasist aspect to Wright is more than apparent in his work.[13] The philosopher Immanuel Kant, reappraising the autodidact Wright, re-interpreted or at least ignored his mystical imagery to derive a disc model, with tighter and more logical tautologies to Wright's theories as what we now know more familiarly as galaxies.[14] Naturally this is not the entire story: more classical mathematical/astronomical appreciations of disc-like galaxies came with for instance Jacobus (Jacob) Kapteyn in the 19th Century. How for instance the spiral shape of our own galactic disc was determined with the aid of American radio telescopes, hydrogen waves, and acoustic-velocity measurements of atomic motions in the 1950s.[15] So then we knew our galaxy was just like those spiral discs seen by Herschel in his telescopes. But this is the metaphysical description nearly always overlooked or forgotten.

So they had the proper perspective now of stars. And stars in swarms that could even be swirling in clouds of gas. This was starting to be assumed about our Sun and its place amid other stars like in the Kant model.

But, what of the energy involving all this gaseous matter?

This is where the story of stars becomes more complex if interesting. Any discussion at all on heat and motion in regard to solar operating mechanisms using calculations was close to miraculous given the time. Early in the 1820s the French mathematical physicist Joseph Fourier was attracted to the idea of the dynamics of heat (energy transfer). Heat flow between molecules is proportional to small temperature variations, he knew. Fourier left an unfinished work on determinate equations, revived by French mathematician Claude-Louis Navier (of Navier-Stokes equation fame). Thomson, later Lord Kelvin, was surely apprised of Napoleon's scientist's ideas and equations. Yet his concrete analysis of the Sun arguably begins with fellow-physicist Julius Robert Mayer (later von Mayer) in wondering what powers a star, in this case, Sol. It was a brave question to ask as early as c. 1848. Such questions still are and, it was all wrong in its details of

---

13   "In the Great Celestial Creation, the Catastrophy of a World, such as ours, or even the total Dissolution of a System of Worlds, may possibly be no more to the Great Author of Nature . . . and in all Probability such final and general Dooms-Days may be as frequent there, as even Birth-Days or Mortality with us upon the Earth." (Ibid, Crowe. Pp. 369-370). Worthy of note is Kepler's similar tendency to do the same fantasizing at times.

14   Kant in his *Allgemeine Naturgeschichte und Theorie des Himmels* (1755) or, *Universal Natural History and Theory of the Heavens*

15   For example, the radio telescopic survey of our galaxy done at Koortwijk in the Netherlands from the Northern Hemisphere and in Australia for the Southern Hemisphere. The survey mathematically and observationally delineated the spiral nature of our home galaxy.

course. However, what mathematics Mayer used was the crucial thing as well as how he applied certain physical laws in the then-kindergarten world of thermodynamics.

It was being implied near this time that sunspots had something to do with the Sun's energy—and even its effects on crops. But the energy source Mayer proposed for the Sun was external: meteorites. Without this energy, the Sun would cool down within 5,000 years he calculated. Unlike Herschel, who was laughed down at the Royal Society for (correctly) linking low-sunspot count years with stock market price-rises due to reduced bushels of corn (that is, wheat) this German was simply ignored regarding his work on heat and motion, perhaps due to such brave speculations as a meteorite-powered Sun. For meteorites and comets do strike the Sun constantly. Such errors are so precious to us in later ages. For, lest we forget in our later times, what was totally wrong then becomes either half or even entirely correct, or at least builds in the direction toward the correct understanding. No immediate genius usually credits science. The whole is a process, often cruel: credit for his work presumably went to another. The Germans awarded him an honorary doctorate and the equivalent of a British lordship—the "von." In any case the kernel of this idea was sewn for the great Kelvin.

The meteoric theory of solar energy lasted a long time, though perhaps not as long as Galileo's on sunspots. It assumed various forms and guises across the mid to late 19th Century. The simple and uninhibited imagination of Galileo talking of a mineral compound that could demonstrate what could be going on in the Sun was overlooked if forgotten: specifically, that some kind of chemical process perhaps *inside* the Sun was responsible for its power. Meteors were more interesting in this regard as an external-energy source since the early dynamicists could hardly imagine a self-powered system constantly fired from within. No known energy producer of such power was then known and it was not to *be* known for a very long time. Mayer's metaphysics were wrong. However the following is *not* wrong, should we heed Kelvin's dicta. We had a quality or a quantity, if you please—the meteorite—with a specific mass and so, the probability of momentum: as proper a training wheel for larger calculations and thoughts as any as to energy produced, dropping into the Sun. It was assumed meteorites thus provided a mass of fuel. All of this was to produce the obvious heat from the energy, thence transferred, say, towards Earth. These thermodynamics pioneers worked such things out, with John Herschel (William's professional-astronomer son) and others already chopping out a rude "solar constant" as to how much solar-transferred energy (or, heat) the Earth's atmosphere took in. (The egg basket of what later came to be and which is still called Total Solar Irradiance—namely, "TSI"—was already being filled.) It was a shame that as regards universities, Göttingen did not discuss matters of (electro) magnetism (Carl Friedrich Gauss' mathematical-physical contribution) in celestial mechanics with Tübingen University and solar dynamics (Mayer). But the then-notion of measuring magnetic fields in different areas of the globe (as was done by a British analogue, General Edward Sabine) being tied together with an inner solar power source was too much to hope for. Like James Clerke Maxwell, Gauss' contribution to the basic mathematical conception of planetary and stellar physics was exceptional. Maxwell's laboratory calculations connected electricity, magnetism and light all to the electromagnetic field. This field was measured in a practical manner by Gauss, on Earth. Sabine of the English admiralty perpetually wondered over these variations of the Earth's magnetic field, disturbing compass readings and perhaps, even the weather?

It was seen that, given meteoritic import of mass that an annual "in-fall" would have been 1/17 millionth of the Sun's mass.[16] As science historian John North put it in effect, even this tiny quantity could be ruled out as being "too great" which was, as he further notes, a sign that the thinking was quite advanced all around howsoever misplaced and incorrect in the embryonic world of celestial mechanics. Kelvin (still Mr. Thomson at the time) demonstrated (1854) that it would mean a shortening of the time of Earth's orbit, were this the case. Hermann von Helmholtz came in with a gravitational energy conversion theory that might lead to knowing the energy transfers involved (the "heat" in other words) better near this time. Such matters as what powered the Sun, and what possible effects it had on Earth, magnetically or otherwise, began what became to be known as the "fifty year's difficulty."

That the optician Joseph von Fraunhofer had first invented, and then used, the spectroscope to see lines in the Sun [17] hinting of chemical processes there as early as the 1810s was not explained by Gustav Kirchhoff (and Robert Bunsen) until the 1860s. That is fifty years between the finding of the one and then, the other, on this subject alone. It was ten years before Fourier's contribution on heat transfer and molecules. It became known in the 1860s—around forty years after its discovery—that the spectroscope showed heated solids and gases making not only light, but also emitting spectral lines of an element at continuous (solid) or at individual (gas) wavelengths. Also that a heated solid with cool gases surrounding it produced light at nearly a continuous spectrum, with blank spaces seen at individual wavelengths. And although Kirchhoff knew of the colors and the emitta, and all its empirical oddities, and listed it for all to see, he and the rest of science had no idea of the *energy levels* of the atoms involved in the elements. Knowledge of energy levels did not come about until Niels Bohr and the basics of quantum mechanics appeared some forty or so years on, further still. Kirchhoff and Bunsen's work led to the aforementioned spectral classification system finally ordered legibly and logically by the Harvard scientist Annie Jump Cannon.

But, to go back to 1860, the discussion of the Sun's operating "mechanisms"—according to laboratory math models of Maxwell and the overly-refined meteor-solar power theory, and perhaps others—lapsed. Considering that by the end of the 1860s, what became known as the Periodic Table of Elements was just being presented in its first tenuous modern form by the Russian Dmitri Mendeleev (that is, for not only the system of classification by groups, number, weights, but also, prediction of *new* elements) it is no surprise that observational work

---

[16] Ibid North, p. 459

[17] Dark line features in the Sun were noted by William Wollaston as early as 1802. Analyzing variations in spectral lines split in stationary magnetic fields (the Zeeman Effect) was not done until the Twentieth Century.

in chemical and element propinquity and dissimilarity was considered alternately frivolous, mysterious and untrustworthy. Mendeleev himself would be six years in his grave before even what he outlined as the strange nature of atomic weight and number could even be somewhat more clearly explained as regarding element classification and the discovery of new ones. And how did Mendeleev factor in to things on Earth—let alone the sun and stars?

Dmitri Mendeleev (1837-1907) Russian creator and explicator of a workable periodic table of elements that would eventually help unite chemistry (matter) and physics (energy).

The difficulty of what powered the Sun from an esoteric as well as practical position becomes very clear. The esoteric questions remain of course as well as many pragmatic ones. The practical value of such knowledge of the Sun was worthless in an age marvelling at enormous steam engines and trains powered by that miracle, coal, zipping past at nearly 40 miles an hour. Energy emission was at a pre-atomic knowledge from 1860-1913 in spite of the advances made in thermodynamics and even in rudimentary nuclear theory (all of it a bagatelle at the time, the Curies included) practical engineers holding on to how much you could burn coal or wood or the newfangled benzene on Earth to get an idea of what could be going on in the Sun, when this odd question arose. The science of thermodynamics is that curious juxtaposition of energy transfer (heat) into mechanical force which grade school science teachers tell you must never be confused with each other especially when you mention gravity in relation to them. Could the Sun contain some unlimited source of a coal-like fuel? This just did not seem real, in spite of all the theory opening up the potential power sources that could or might perhaps exist. The knowledge of a nuclear fissile power was just as strange and unknown then as being able to control nuclear fusion power in an enclosed environment on Earth is today.

Kelvin was the brave, emergent thinker here in the mid-1800s. By 1854 he equated heat with practical, process "work" in a way that each and every unit rise in temperature meant a parallel increase in work accomplished. An absolute temperature scale that now bears his name is one of his contributions in calculating mechanical or electromechanical "work" involving heat.

Seeing that in 1854 Ernest Rutherford—arguably the father of nuclear theory—was not to be born for another sixteen years, this was sensible. Eighteen fifty four was perhaps the pivotal year of Kelvin's dismissing the energetic force of the Sun in ways that became much clearer by 1892 and extremely evident by 1912 and in the 1930s. Coming to loggerheads on how the Sun effects the Earth in energy transfer, or how it operated at all, given observations, was an aim at ending the "fifty year difficulty." This presumably was solved by Edward (E) Walter Maunder. Kelvin, who acknowledged the difficulty, probably never accepted it even if Maunder publically proved it. No data on his response to Maunder is recorded and Maunder refuted Kelvin unchallenged by the great man. From a slew of people, Maxwell, to Count Rumford to N. Sadi Carnot; from Rudolph Clausius to the brains-in-his-fingers Michael Faraday, James Watt, and even perhaps to the redoubtable Mayer, Kelvin knew that two key laws of thermodynamics work together. The first to conserve energy, and the second to lose it (entropy). Since it could be assumed that any cooler area (say, theoretically, an area around the Sun that could be seen as being very cold [18]) these two laws, taken in tandem, required an input of energy from outside the system to maintain its energy-emitting or transferring force. It was the same thing Mayer knew. Given such logical insight, it is no surprise that the crashing-meteorite-into-the-Sun energy-providing process made good sense to these scientists then. It was no wonder that Kelvin, honest, master calculator that he was, could only come up with some 3,000 years of ultimate solar power in the long run in one calculation.

The scientific imagination starts to run thin and the numbers and calculations too thick concerning Sol—though the data on the Sun increased around 1860 and enormously so after 1870. By 1903 Maunder had a thirty-year photographic database of it at the Royal Observatory in Greenwich. The then-known power sources (coal being paramount, petroleum just appearing) could not possibly produce enough energy to power something with so much mass—the Sun even today calculated as comprising nearly 99% of our Solar System. Unsurprisingly Kelvin was never convinced. As far as he could see, no known chemical reaction (should such even be possible in the Sun) from any existing power source could radiate it out for very long. Being the re-formulator of Maxwell, the most practical research-to-applied engineer then living in a world that craved technical marvels and instruments of ease more than we do, Kelvin's word became final on most of this subject. His word followed English imperial power especially to a financially docile and receptive United States. His mind was firmly entrenched in the very concrete aspects of linearity in the Sun's probable energy production and transfer and the attendant lack of power it must have upon the Earth due to all this. Other personal reasons for Kelvin denying an active, non-isotropic Sun will not be speculated upon in these pages. Yet the force of his massive influence cannot under the circumstances of his time be denied, for the good or the bad.

---

[18]   Although we now know that the area beyond the Sun by a great distance is actually hotter than the solar surface itself, put yourself in a physicist's shoes c. the year 1855.

Much later on, there was a disconnect between the nuclear nature of the Sun, as this became known through the Twentieth Century quite clearly, and its ability to be non-linear. Arthur Eddington, reviewing Subrahmanyan Chandrasekhar's purely mathematical descriptions of maximum masses of dwarf stars justifiably held Chandrasekhar's work at arm's length although in this case, the latter would prove to be profoundly correct. For not all mathematical predictions and assumptions of the ongoings in Nature are simple quantitative speculation. Although Eddington's assault on the great Indian physicist was justified in that no observational proof was forthcoming from Nature at the time for it, crucial love was lost between these scientific powerhouses.[19] As such, the emergent knowledge of our Sun being nuclear and very unstable, much as other stars, became even more publically buried. As battles like this raged in the 1920s and 1930s, a constant and very linear Sun became the standard textbook description: maybe even folk wisdom in a way. Even less could most people picture the Sun as something not only powered from within by some awful source, but that it could work on and off, and in an unbalanced, almost capricious way we term for convenience' sake non-linear (or non-isotropic). Old justifications came to the fore, and lassitude in understanding grew paramount. And yet as early as 1903 some had not only seen this behavior by the Sun, they could also mathematically delineate some of its then-inexplicable non-isotropic flailings and lashings—regardless of whatever it was powering it. Earlier still in the 1840s, some noted a connection between geomagnetic disturbances and sunspot patterns, and much later in 1909, connections between Indian cyclones and solar rotation from data gathered in the 1850s. This "disconnect," you could label it, has lasted down to the current time in some crucial and stubborn ways.

We have just seen how the natural lapses in scientific discovery and explanation can *conceal* regarding this most important of celestial objects, our Sun. Thus lies the challenge to see past enforcing an order on it that is untenable due to self-imposed or arbitrary limitations in metaphysics, mathematics, communication, and useful speculation and to thus find out just how it *does* operate in all its nonlinearity. If not this, then any partial understanding of operational stellar nonlinearity especially in relation to the Earth is preferable to none at all.

---

[19]   Some scientists (in the best tradition of science) insist almost absolutely on observation matching the mathematical or other prediction. Eddington himself had gone to great lengths to show Einstein's prediction of the curving nature of light in 1919. Sometimes such scientists can be harsh in their criticism but justified.

(Left to right) Arthur Eddington and Subrahmanyan Chandrasekhar

Entertaining dynamism in the Sun of most kinds was held suspect and even ignored, it is said, and is sometimes baldly shown, for many years after Kelvin's sway took hold in the textbooks and in the physics and chemistry classes. Great men do great things and sometimes perpetuate great mistakes quite unbeknownst to them and their intended benevolent auspices. They are not to be blamed. Solar dynamics as we now know it was not really to gain a footing toward a working theory until after World War II. Given the lapses in discovery and the slower take-up when these discoveries were acknowledged and connected (and reconnected) and the mentalities of the respective ages, their limits and glories, it is of no great surprise. But dynamic the Sun most assuredly is, no matter which personage said what to the contrary.

◊◊

This book is about how the solar dynamo theory from some research is being used to explain certain peculiar and usually inexplicable behavior on the Sun that shows potential for understanding useful, repeatable aspects in the Sun. Much of it is wrong or half-right; some is probably entirely correct for now. Unlike other plausible theories of how the Sun functions, the solar dynamo theory [20] is not only a predominating theory based on classical physics. It has been quantified with hard mathematics. The "solar dynamo theory," as it is called was first advanced by H. W. Babcock in 1961 [21] and later refined by R.B. Leighton [22] (1964; 1969). These names are linked together as "Babcock-Leighton models" (mostly conceptual/ "hard" math, not statistical) and are vital in explicating the theory in different ways. An even more mathematically developed scenario for the solar dynamo was by Max Steenbeck and Fritz Krause (1967) and

[20]   Like Darwin's Theory of Evolution, and Newton's Theory of Universal Gravitation, and Einstein's Theory of General Relativity, the "Solar Dynamo Theory" reigns currently in the "most workable theories" department regarding the Sun. It has, unlike the other three still-to-be-disproven theories, no single author and is definitely "a work in progress."

[21]   Babcock, H.W., 1961, The topology of the sun's magnetic field and the 22-year cycle, *Astrophys. J.*, 133, 572-587

[22]   Leighton, R.B., 1964, Transport of magnetic fields on the sun, *Astrophys. J.*, 140, 1547-1562. and later, Leighton, R.B., 1969, A magneto-kinematic model of the solar cycle, *Astrophys. J.*, 156, 1-26

Steenbeck et al. (1967).[23] These gentlemen presented some essential elements for the theory as proposed earlier by Eugene N. Parker (1955).[24] Steenbeck had been at the German Academy of Science's institute for Magnetohydrodynamics (MHD): one of those talented German physical science chess pieces moved dangerously around by German, Russian, and American political and science bureaucrats throughout and after World War II until the middle of the Cold War (1945-1991). Some of the other input into the theory comes from before World War II in the pioneering work of Sidney Chapman and Julius Bartels, Chapman actually having worked with and for E. Walter Maunder and his elite mathematician wife, Annie, at the old Royal Observatory at Greenwich, England (during World War I).[25]

One thrust of the solar dynamo theory is, then, that the Sun's activity and its variations are assumed to be driven by a magnetohydrodynamic dynamo. Magnetofluiddynamics, or hydromagnetics, is the academic discipline which is the study of the dynamics of electrically conducting fluids. Now we come to Navier-Stokes. Besides Babcock-Leighton models, we have MHD using "Navier-Stokes" equations from the 18th Century to describe the motion of fluid substances.

The Sun is one, gigantic fluid ball that is electrically conducted, in other words.

Thus, the solar dynamo as a theory about the Sun deals essentially in electrically-conducting fluids and their dynamics and how currents are induced in these fluids, and which trace the currents' motion.[26] These phenomena are described mathematically as far as is possible, and all is driven by the power afforded mostly by what after World War II came to known as the Sun's nuclear fusion core "reactor." This was the elusive internal solar power source that baffled the 19th Century thermodynamics researchers, and which challenged some even later on in the Twentieth. First, consequently, is involved in the dynamo—electromagnetic force. Very much of it.

It is like taking an electrical lab for all this, using what Michael Faraday, Maxwell, Kelvin, Babcock, Leighton, Fourier, Navier, Krause, Parker and a hundred unnamed Swedes and Russians etc., threshed out, then seeing if it can be applied to a piece of swirling, seething nuclear-forced Nature that is about 800,000 miles in quasi-diameter with spots popping out over it at times in apparent stringy loops, when bright.

We'll never get all of it right. But to get some of it right could be very useful, indeed.

---

[23] Probably Krause F., Steenbeck M., 1967, *Z. Naturf.*, 22a, 671. Possibly Steenbeck M., Krause F., Rädler, K. H. 1966, *Z. Naturf.*, 21a, 369. The work was most likely about turbulent plasmas.

[24] Parker, E.N., 1955, Hydromagnetic dynamo models, *Astrophys. J.*, 122, 293-314

[25] Chapman and Bartels' work on geomagnetism resulted in a work on the subject entitled *Geomagnetism*.

[26] Helioseismology (to be explained later) has an acoustical element in tandem with this motion.

◊◊◊

The current and coming world generation of aerospace engineers, geoscientists and astronauts, armed with all workable science, and the best paramilitarily trained, look up at the Moon and at Mars. What lies there for humanity? We see on the face of an explored Moon a reason for making an astro base, thenceforth a possible launching pad for routine Mars missions. Many of these missions will fail. Lives will be lost. The Sun stands by and in the way: sometimes helpful and sometime not. The "not" part must be understood

Why must we do this? There are, superficially, two good reasons among many. One is for mineral exploitation and human residence. Several billion humans need more room. This is engineering and simple common sense. Knowing how the Sun functions as our particular variable star facilitates this exploration and settlement and secures our better well being on Earth. The other is simpler still, on the one hand, but ultimately, like the "soft sciences" of sociology and anthropology, more complex than the "hard" physical sciences on the other as E.O. Wilson put it once, to effect. We must show ourselves that we can do it. Humans will to know. They also will to dare. We saw in the Moon landings that walking on planet-like objects can be done.

At NASA a number of ambitious and realistic research and test projects aimed at just such exploration and possible habitation of nearby satellites and planets is ongoing. Getting used to existing in space for long time periods is one major goal. Such things taken for granted on Earth are major obstacles in space, to include behaviorial adaptation, conduct control and management, and the even mundane but very important need for exercise. NASA's Living and Working In Space (LAWIS) project has been exploring avenues by which to come to terms with renewable and sustaining ecosystem management with an aim towards self-reliance on the Moon as well as possibly Mars. A knowledge of how the Sun functions goes a long way toward ensuring not only the physical powering of spacecraft, but also how to protect persons in long time periods from hyper solar activity, the watching for warning signs of unpredictable solar activity, and how to thrive on reduced amounts of the same. Questions as to how effective solar power would be on the Moon and Mars (the latter to facilitate crop growth, say) are also soon-to-be coming items of interest. Technology has long since outstripped the tests done in Earth-based closed habitation systems like Biosphere I and II of the 1970s. How do cosmic rays affect climate on Mars as opposed to Earth? Is there a way to make a proto-climate with technology using the Sun as a power source? Would solar heating up there be the best and most effective means of facilitating long-term enclosure existence? If so, how? What about a better understanding of how the Sun functions to enhance and make redundant even Earth-based solar power for mass-market electricity creation? Will extended solar grand phases for the higher or lower have fortunate or deleterious effects on human-collected solar power on Earth, or do they matter in the long run for such purposes at all?

Companies like the relational database firm Oracle sponsors the Oracle Education Foundation to stimulate young peoples' interest in planetary study and potential human exploration and eventual settlement, such as the Mars Academy. Foundations such as the Mars Society have an international reach to support space agencies in the direct settlement of Mars. Knowing better how the Sun functions between here and there will be absolutely vital for any such endeavor—even for success and comfort in the much more mundane and relatively easily-achievable Moon base programs for mining and other interests. But we cannot wait until then. Space missions such as Ulysses, Advanced Composition Explorer (ACE) Yohkoh, the Solar and Heliospheric Observatory (SOHO) and Transition Region and Coronal Explorer (TRACE) have and still do contribute to increased data collection and knowledge of the Sun. The latest effort, the Solar Dynamics Observatory (SDO) is collecting data at this very moment for interpretation. The SDO is a space-based observatory working far above our atmospheric-muddled view of Sol. It is functioning on a close, continuous, and three-fold manner gathering information much more effectively than ever before in some crucial aspects that will, coincidentally, be the over-arching theme of this book. Who will interpret the data? What conclusions will be reached?

Four thinkers who helped bring forth the modern technological age as guided by science, nearly all of whom had significant, insightful—and imaginative—views concerning how the Sun works. (Right to left, Frances Bacon, Rene Descartes, Galileo, Johannes Kepler)

Most important of all, who will use their scientific imaginations best to come up with newer insights into how the Sun works?

Although the Sun is neither the center of the Milky Way Galaxy and is a most nonlinear and aberrant entity to some ways of thinking, expanding on some aspects of its repeatable nature are an absolute must if we are to mine and settle other areas of the Solar System.

We must also know what it does in order to guarantee the safety of power and fuel lines and satellites in an incredibly interconnected age. Perhaps such a knowledge will guard us one day in being able to predict future long-term climate change.

1. Comprehending, and contemplating, deep time

Isotopes are vital bits of proof for otherwise invisible connections in Nature (and hence, us *in* Nature) that have lapsed as school and general knowledge. Knowledge of isotopes never gained strongly amid the general populace. This is due perhaps to its multifaceted nature. They are complex to understand even when put in the simplest terms, are used by specialist scientists, and have only fairly recently been considered effective in use after a trial of many years since formulation. Yet very modern science literacy depends on knowing what they are.

Father of evolution Charles Darwin had no idea of what isotopes were, having died long before 1913 when the process was presumably named. But as a geologist, just looking at the sheer length and breadth of some life form developments, branching-offs, dead-ends, and re-emergences in altered if similar forms from stone and in life convinced Darwin—a bad mathematician if there ever was one [27]—that Earth's age must be vast. He came to this after a long lifetime of continual analysis and re-analysis of fossil, erosion, and other factors. Poor in using discrete, thin mathematical symbols and algorithms to compare vast quantities, he had, on the other hand, the time and energy to compare vast quantities from their concrete (non-mathematical) substance using conceptual reasoning. So of course he had no concrete proof, numerically (and mark this word) for the great Earth ages fossils apparently had. Later, after the discovery of the isotope and its fitting in to the framework of deep time, he was to be proven correct in terms of Earth's vast age more than even he could ever have imagined. [28] It leaves open for the unadjusted and disinterested scholar the vast "mind scapes" (and intellectual "mind fields") in the contemplation of just exactly where modern humans find themselves, and from where they partly came.

---

[27] Recommended here are the books of Stephen Jay Gould, some of which uncovers biographical detail regarding this scientist as well as discuss his method of doing science.

[28] Darwin had also pieced together inheritance of traits in mosaics and branches in descending life that were later more tidily accounted for by the monk, Gregor Mendel's, work—whom he knew nothing of over in Austria, but whose work was disinterred years later, thankfully, because that was the root of what we now call genetics, and the medical view of our inherited physical traits. The value of this science goes without mentioning.

(Left to right) Charles Darwin and William Thomson (Lord Kelvin) in later years.

Cautious for fear of being wrong most likely due to a greater fear of numbers, Darwin found out that the great ages he suspected of fossils "could not be" anyway, since a master mathematical physicist had "proven that it could not be so." In the latter's case, this was due to numbers. That is, numerically, a deep age for Earth could not be proven and based on the physical concepts as they were "then known to be," and as he (the mathematical physicist in question) commanded them, by reputation, to be known.

Here arises a common problem with utility, persons of great utility, acquiring scientific knowledge, the profit motive and its motivators—and lastly, human nature. We meet him here again and will not judge, but examine, and ponder, his thoughts and acts. William Thomson, later Lord Kelvin (Baron of Largs) was a brilliant mathematical physicist with, as in most utility-minded engineers of the best sort, a tendency to limit his abstract thinking to increase the strength of his detective work in finding out "why?" and "how" in very practical problems (such as in instant communication, or as in cheap lighting, or as in more efficient machines). This is a metaphysical thinning process: not a metaphysical thickening process. This thinning process was—and is—of incredibly great value to setting up and making refinements to practical things for common utility based on known quantities and aspects. For example at the time, the trans-Atlantic cable, connecting American and English stock markets in seconds instead of in days or weeks. It was priceless insofar as Kelvin recognized the value of Nicola Tesla's AC [29] generator over the extant Edison DC variant that was inefficient (leading to high costs as well as to terrific danger and pollution) so that Westinghouse brought us that.[30]

---

[29] Nicola Tesla claimed he had a "vision" of how alternating current works by looking at the Sun. See *Tesla, Master of Lightning* by Margaret Cheney and Robert Uth (Barnes and Noble: 1999). Tesla was an Americanized Serbian.

[30] Alternating current can conduct energy over long distances, increasing energy conservation, circumventing entropy, and so chopping cost whilst boosting availability. AC has been credited with "inventing" the

Kelvin however was not all "practical" engineer *per se,* He truly was a brilliant theoretical scientist. He devised the concept of absolute temperature (absolute zero) for which he is named (the units called Kelvins) and for which deep-space knowledge of temperatures would not be possible to comprehend. He revised thermodynamics, period—perhaps modernized aspects of it for all time. He became wealthy due to some of these and other projects, and was lionized and lauded (knighted as Lord Kelvin). He became the Secretary of the Royal Society of England. The influence of this society is enormous to this day, the Secretary's [31] chair having been filled in the past by personages no less imposing than Isaac Newton (who wielded a similar power, similarly).

Knowing as already discussed what might fuel the Sun at the time, Kelvin showed why the Sun could not be so very old from concrete facts as then known. The fuels he knew of could not possibly burn for any more than a few thousand years.[32] The calculations from internal burning hydrogen led his own calculations to show that the Sun was, thinking from this angle, less than 50,000 years of age and earlier, we saw another number that was probably then even more well accepted: 3,000 years. He had thought through external "heat" transferring objects affecting the Sun (as mentioned) as the early 1850s, perhaps earlier. He also calculated the age of the Sun in visible light (not much known about the invisible forms then.) Then he:

suggested that the Sun might have formed from a giant gas cloud. Gravity would eventually cause the cloud to collapse into a ball, since each molecule attracts all the other molecules. As with any falling mass, the potential energy of molecules in the cloud would turn into kinetic energy as they fall. Kelvin reasoned that this energy would turn into heat . . . While each molecule would not add very much heat, there were a lot of molecules! Kelvin was the first to suggest that the stars form in this manner. *Though we now know this is not how they generate all their energy,*[33] this is how we think stars initially ignite the fusion process. Today's astronomers know this mechanism as "Kelvin-Helmholtz Contraction." (The German scientist Hermann von Helmholtz had independently proposed a similar theory.) Kelvin concluded that the Sun built up all its heat as it formed. During its life, it radiated that heat away like a hot coal. He estimated the lifespan of the Sun to be about 30 million years. This remained the standard solar model among physicists and astronomers until the early 20th century.[34]

Stunning as this achievement was in the annals of science, done by one of the greatest living scientists, it is limited and wrong in many parts. (It is OK: it has only to be falsified.) Additionally, some of this thinking persists into the 21st Century in terms of the Sun's linearity—radiating

---

modern world.

[31]   Known as President as well

[32]   *The Age of the Sun: Kelvin vs. Darwin,* by S. Gavin, J. Conn, and S. P. Karrer. Physics Department, Wayne State University, Detroit, MI, 48201 sean@physics.wayne.edu

[33]   Italics by the author.

[34]   Ibid, Gavin *et al*

heat (rather, transferring energy) actually all its energy and light away in all directions, at once, steadily. Kelvin's work has been the sine qua non of solar science ever since. It is the shining hypothesis that remains to be falsified in all its parts. Of course it has not and might not ever be. But still, the greatest challenge we have facing this is that the Sun is less linear (very much more non-linear) than many wish to think.

Rolling around the figures from various calculations, Kelvin took the Earth's age to be as high as 400 million years—which could NOT have been younger than the Sun. He settled in the 20-40 million year mark. He was a king of numbers, thin concepts, and the quiddities of either not to be crossed, his contributions to applied science creating enormous personal and societal wealth. He was religious and could have been chary of his numbers, not wanting to contravene accepted religious wisdom and Christian scripture (Darwin was by far the less conventional figure here). Aging, even then-newer theories emerging on radioactivity (Rutherford for example, whom he attacked and denied) and on thermal conductivity could not sway Kelvin towards the Earth's probably terrifically deep age, and his imposing mien and reputation drowned out much. The Victorian manners of the many clapped hands and nodded aye. In 1885, Kelvin utterly rejected any dynamism in the Sun and denied any magnetism by sunspots. By 1903, when challenged by the retired observational and photographic solar researcher of 30 years at the Royal Observatory E. Walter Maunder as to the most likely very non-linear nature of the Sun, to use many modernized concepts not devised by Maunder, Kelvin dissented again, and yet was proven wrong at the Royal Astronomical Society by Maunder in person without Kelvin's attendance to disturb the peace. For math models to buttress Maunder's delineation of the observations (after 30 years of constantly recording and attempting to interpret the Sun's behavior) were provided by Maunder's second wife, Annie Russell Maunder. Kelvin was not on hand to respond to Maunder and never did. [35]

◊◊

After the early thermodynamicists, ideas and tests formed by the likes of Marie Curie and Ernest Rutherford grew stronger. By the earlier part of the last century (Kelvin was physically dead by then: his reputation as influential as ever) Rutherford's and Curie's and others' thought and experiment burst forth as if from a chrysalis and thus floated, butterfly-like, above heads wanting enlightenment. The beauty of the science was clear. It was found that when Mendeleev's elements' atomic numbers are changed due to an atomic mass change in an element, the neutron amount differs so that a total proton-neutron count for that element's nucleus alters. The element Beryllium for instance becomes something different. It becomes an isotope [36] of Beryllium ("Be" is the Periodic Table "symbol" for the element, Beryllium). The notation becomes different, as well. It was the missing link that Mendeleev sought in that table of elements scientists struggled with since the days of John Dalton and Amadeo Avogadro. If

---

[35]   Yaskell, S.H., *American Scientist* (July-August 2008), "Mistakes were Made" (Letter)

[36]   The name for this was given by a woman.

the new atomic mass for what was (formerly) Beryllium is now 10, then the notation is Be-10 (or 10Be: notation varies) instead of simply "Be" to denote it as something other than the original element (the element Beryllium ordinarily has an atomic mass of nine).

Such things were discovered working with and studying, for example, the element Radium (tissue-destruction and "alpha" and "beta" waves also were found from researching Radium). But without others in the then-nascent field of nuclear physics from 1890-1930, such as Ernest Rutherford (electron-proton model with nucleus, proton in the nucleus, electrons in the rings) and James Chadwick (who added neutrons to the nucleus of Rutherford's model, so that we could discern mass better in Nature with its all its changes) it would have been unknown. The very instability of Radium's nucleus showed the hops and jumps in mass, etc. Radium presents for us a very nice working example of an element in its flexibility.

Niels Bohr near this time saw the workings of quantum force (force as thermodynamics) on an element's atomic makeup. This left not only Kelvin but even Rutherford in the dust. Where Rutherford had been simply atomic, Bohr delineated a strange but mathematically-obedient *sub*-atomic, world. Many electrons are fused as it were to a "shell" in the tiny Bohr world of quanta. They then follow strict rules in their separate rings which sometimes cannot take another electron from anywhere else. If energizing from any source causes enough of an excitation in an atom, then these electrons can be belted around to change series states of energy (as in Balmer, Paschen, etc.) and they then release photons—which are blasts of light energy. The discovery of the positron by C.D. Anderson to fix the ever-chronic mass trouble, being the opposite as it were of the electron, led to understanding what a "cosmic ray" really is (a real sci-fi term if there ever was one: it's not a "ray," and "cosmic" is too vague).

A cosmic ray is best described as *a super energetic force from nuclear bodies*. Some of these nuclear bodies are known as stars. One of the stars doing this with rays is *Sol* . . . another word, a Latin one, for our Sun. So as Mayer and Kelvin could not fathom due to no fault of their own, the Sun *was* internally powered Yet this still is not to say that it *cannot* be externally influenced by other, stronger forces. For there are other forces stronger than our sun.

What is known about "ray" (say he is a tough, good-looking Catholic kid from Brooklyn) is that when "Ray" (as in cosmic ray) gets punchy after Sol throws him out of the his deli for being a wise guy, strikes an atom of an element, the atom is shattered. The particles from it speed off in all directions. If the element loses an electron it becomes an ion in this process (hence "ionized gas," like when elements' particles cosmogenically get forced into a gas this way in the upper atmosphere of Earth by the Sun's action). A lot of these ions then release light (photons) if electrons are excited up to larger distances from the atomic nucleus and then fall back, thus giving rise to various series-steps of radiation (say Lyman, Balmer or Paschen [37] Series). Anyway,

---

[37]   For T. Lyman, J.J. Balmer and F. Paschen

if for another strange reason the atomic mass (the proton-to-neutron ratio) in a particular element's nucleus is altered due to such a force, it becomes what we said.

An isotope.

Isotopes of elements are made all the time between us and our metallic sun (and other elements on the Periodic Table) at all kinds of strengths, times, and speeds in the upper to lower-Earth atmosphere. It was the missing link you could say in the understanding Kirchhoff had arranged and sought in the spectroscopic analysis of the Sun. In any case these "cosmic rays"—super energetic forces from hyperactive nuclear bodies—can now due to experiment account for assisting in making cloud condensation "nuclei" [38] that starts when atoms are shifted around by these forces, and can help form clouds one mile up from Earth's surface.

But, why would they do that? Is it that cosmic rays come from the Sun and other stars at us all the time, steadily, and sort of politely? That answer is no.

Cosmic rays may indeed come from the Sun at us more or less constantly if not evenly and all at once, at equal rates. That is, when they come, they come at us like the Sun itself does: *non-linearly.* They may also come at us on Earth all the way from some exploding star in our very own galaxy (a particular kind of supernova) so far away in absolute magnitude we cannot see their radiant energies. But their "force" (like in Star Wars) is with us or coming at us, strongly or weakly, as energy, depending on how close they are in relation to us. Also, two other things will "handle" ("modulate") these "rays" in relation to us on Earth:

1. The Sun's "magnetic sheath"—actually itself a magnetized cloud of ionized gas surrounding the Sun up to distances as far as many hundreds of times the distance of Earth to the Sun (some call it the Parker Spiral)
2. The Earth's magnetic sheath (magnetosphere, which several other planets also have, Venus having a very weak one).

If the Sun's magnetic sheath is flabby due to a non-linear reduction in its nuclear power generation for x amount of time, Ray will come in trying to kick Sol with a vengeance, Sol perhaps releasing weakened forms of Ray. If Sol's magnetic sheath is flabby, Earth's is the same, since "if he's pusillanimous so aren't we." All ionized gas and isotopes obtained in our atmosphere's ionosphere from the Sun or these supernovae (if in our path)—or from both—find their ways straight down into the Earth's lower atmosphere via the ionosphere, trickling down, so to speak,

Some of the material "trickling down" is the isotope Carbon-14 (atomic mass changed to 14 from 12). When it is formed, it is immediately oxidized to $14CO_2$. It remains floating in the

---

[38]   Provisionally "proven" at CERN in August, 2011. See Kirkby *et al.* in later chapters.

atmosphere and does not fall to the ground. There is an uptake by plants of 14CO2 right away however. Five thousand (or so) years is the radiation decay time of this isotope, A.K.A (Also Known As) "its half life." The half life is the period of time it takes for a substance undergoing decay to decrease by half, and this is determined by mathematical functions. This is what gives us such an enormous view into the past: those little numbers in the functions attached to this physical phenomenon. Carbon-14 actually decays (transmutes is perhaps a better word) into an isotope of Nitrogen. After 60,000 years, that is it. You can't date things with Carbon-14 anymore. But it can "date" enough: like sequoia trees and pollen samples going back hundreds of solar cycles. Sixty thousand years of them in fact. That is, all those trees (like plants) that live (lived) under the Sun. But carbon perhaps is the strangest of all elements. The Carbon atom is tetravalent, meaning it has four electrons ready for covalent bonding at any time. That means it combines with or becomes all kinds of compounds (like Carbon Dioxide) and can be released (Carbon Dioxide by cement production, say). Carbon has one electron "shell" that can never be filled. Astronomers Robert Jastrow and Malcolm Thompson said thusly about Carbon, speaking sub-atomically:

When four electrons are placed in the second shell of the Carbon atom, they make up, with the two electrons in the first shell, a total of six electrons. Six electrons exactly cancel the positive charges on the six protons in the Carbon nucleus. Thus an additional electron passing by feels no attraction toward the Carbon atom and cannot be drawn into an orbit in the second shell, even though there is a place for it there [39]

as it follows Bohr's and others' odd rules at sub-atomic levels. In any case, if you want to toy around with a physical science substance for abstract weirdness, Carbon is definitely "it." All of life is carbon. An entire branch of chemistry (organic) is all about carbon, so big since it helps us carbon-based human types with our studies for medicine, genetics, and useful (as well as harmful) chemicals we come into contact with daily.

Half of a given mass of Uranium-235, the numbers tell us, vanishes in a similar way after 700 million years (that is, the half life of this isotope of Uranium) until it becomes Lead-206 (just like 14C eventually becomes Nitrogen-14). Finding this stuff in fossils with (the isotope of) Lead ratios really gives us a view into deep time, indeed. Isotopes like this give us an extraordinarily deep view into the origin of life, as the sediment the former life forms are found in contain such isotopes and hence the sediment the life form has become—giving proof they were alive close to or at the time that isotope was created. And so we know that the sun has powered, if indeed not begun, all life as we know it for at least two and a half billion years.

Kelvin would have been amazed by all of this, if not a bit frightened and Darwin, oddly justified and equally spellbound and jarred, most likely.

[39]   Jastrow, R., and Thompson, M.L., *Astronomy: Fundamentals and Frontiers*, 2nd ed., (Wiley: 1974) p. 82.

Without knowing the half life of Uranium-235 the Geological Time Scale would have as much relevance scientifically in terms of its numerical proof for life having existed 450 million years ago, based upon fossil and sediment evidence, as the Bible's figures do of the creation and of the flood. [40]

But scientific texts are not sacred ones. The scientist Robert Boyle, author of Boyle's Law and theological scholar as well would have perhaps been the first to point this out at this juncture. What we are as spirit is one thing. What we are as part of often grim Nature is altogether another. We use science to understand ourselves in Nature. This has helped us to develop things like jet propulsion aircraft to imitate the bird. Also, internally-used anti-bacterial medications to stave off hungry germs to facilitate the feeding of a hungrier fungus we then pass. We consult holy men to get a grip on our spirit. This helps us to live in Nature that we cannot only not avoid, but indeed, are a part of, and often have difficulty understanding. The proof that science is an outgrowth of true faith is the fact that we use science to help us understand Nature which in turn, usually benefits us greatly.

But scientific texts and sacred ones run parallel to, not against, each other. To make it even more mysterious, arguably the first truly great "modern" systematic philosopher of science, Bernard (Baruch, alternately Benedictus de) Spinoza once asserted: God is Nature (and alternately, and without apparent explanation) Nature is God. [41] Spinoza, probably the first published proponent of Nature as a kind of book we can try translate for our well being in all its parts, and the metaphysician who stood at the threshold of what later became known as physics, authored the famous phrase "matter is neither created nor destroyed." [42]

All said, we have been allowed or have been lucky enough to go about our business mutually enriched in this parallel fashion. It has led to a population of several billions of human souls. Many among us are fed and reasonably well clothed, sheltered and healthy. Many are not, but hope is still there, perhaps more than ever before.

For people like Darwin and Kelvin this would have come as a surprise, received probably in amazed and possibly pleased silence: *science has elevated human understanding.*

---

[40] This is not to say that theological numbers are erroneous or fraudulent: they have another intent and interpretations vary. For a good treatment of this see Gerald L. Schroeder's *The Science of God: The Convergence of Scientific and Biblical Wisdom* (The Free Press:1997).

[41] *Deus sive natura, natura sive deus.* See Stuart Hampshire, *Spinoza and Spinozism* (Oxford University Press: 2005)

[42] Originally in a letter to the Royal Society's Henry Oldenburg

(Left to right) Robert Boyle and Bernard Spinoza.

To Spinoza and Boyle it would have been the sign of a miracle.

2. What is a grand solar phase versus a "regular" solar phase?

Some have heard of "solar maxima" and "solar minima," these being the two still-recondite terms for phenomena following a more or less eleven Schwabe (and double, twenty-two year magnetic Hale) cyclical "flip." When one Schwabe "cycle" ends, another one begins, and this period, the lead up to the big show—the solar maximum—(for example, the 2013 solar maxima that could surpass those of the latest cycle) gives us the opposite of maxima. That is, solar minima.

So what are "grand" phases in these regards?

Grand phases are longer-term variations in solar activity either for the stronger, "normal" minima/maxima "flux" of total solar irradiance (GRAND MAXIMA). Or they can be grand phases for the weaker normal minima/maxima flux (GRAND MINIMA episode). These periods have been tied to many cycles, some well-known, and many, not known well at all. When and how long either kind of grand period lasts always leaves ample room for either conjecture or confusion.

As the "normal" minimas and maximas influence Earth climate, so do grand minimas and maximas. Grand phases influence climate and geomagnetic storms that effect Earth in ways that go beyond normal understanding, geomagnetic effects far better recorded and understood than those of climate. It is all still very strange. Grand phases for the higher have gone hand-in-hand with relatively long-term warm periods in the Northern Hemisphere (for hundreds of years, like the Medieval Warm Period, the very existence of which has been debated [43]). Lower ones have coincided with degraded weather in the Northern Hemisphere—like the so-called Maunder Minimum some people have noted and the existence of which is not debated. Why also do we discuss only [44] the Earth's northern hemisphere regarding Earth climate and geomagnetic influence? It is because this is that portion of the Earth so prone to glacial recrudescence (re-emergence) that comes fast and hard over the landmasses where human society has evolved and perhaps has thrived best. This portion of Earth sees the changes in a most pronounced way: whether for a fairly sudden increase in the tree-line north, to a fairly sudden increase in warm sailing weather, east and west, below for example the North Atlantic Ocean in the eastern Laurentian (versus the western Bering) side of the North American continent. Fast in this sense could mean a few hundred years of course.

---

[43]   For instance by Dr. Michael Mann

[44]   Actually, in long-term minimas and maximas, "grand" solar phases, the Southern Hemisphere is also significantly affected.

Let's illustrate what a typical solar minima and maxima is, versus a probably not-typical grand solar minima and grand solar maxima. Curves and amplitude waves demonstrate these as if one spoke into recording device, thence to watch the monitor show loops rise and fall. The amplitude of the waves (all these are waves, and their little ups and downs are the amplitudes of the waves in Figure 1) show peaks (solar maximas) and valleys (solar minimas). The "normal" variant of maximas and minimas in the middle of Figure 1 shows a longer base amplitude (the thick black line). The loops between the peaks and the valleys are wider, freer, in the "normal" or "regular," minima-maxima strophes. This is what could be described as what we have been living through on Earth since about 1724 [45] and at least until 1924.

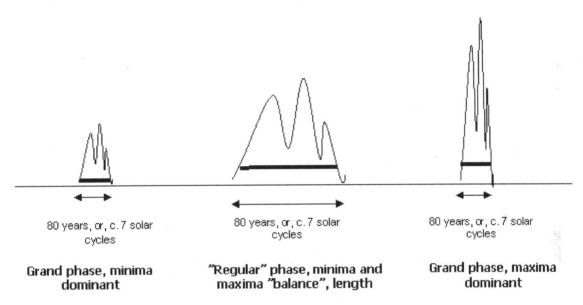

| 80 years, or, c. 7 solar cycles | 80 years, or, c. 7 solar cycles | 80 years, or, c. 7 solar cycles |

Grand phase, minima dominant   **"Regular" phase, minima and maxima "balance", length**   **Grand phase, maxima dominant**

Figure 1. Solar minima and maxima in grand and in regular "phases."

Look at the two so-called "grand" phases that hem the "regular" one in Figure 1. The one to the left has an overall lower amplitude in its peaks and valleys. The one to the right has an overall higher amplitude in its peaks and its valleys. But note that those on either side could [46] have the same base length. In any case, the base lengths in the grand phases are shorter than in those of "regular" periods.

Much more on this will be described in later chapters, but the essential paradigm drawn up in Figure 1 of this chapter will be discussed throughout the book.

---

[45]   This was somewhat before measuring of sunspot cycles began in a consistent manner. Cycle 1 began in 1755, as per human recording. The current cycle is 24.

[46]   It is not certain if the cycles have the same base length in grand maxima as in grand minima. This matter requires further investigation.

◊

The hard-to-believe-and-understand-thing here as regards extended minima is that, what is now called MAUNDER type grand minima was discovered around three hundred years *after* it actually happened. The dynamic of an extended maxima is the logical analogue of the phenomenon of any extended minimum, and though contra one another, are probably part of the same dynamic.

We begin, mundanely enough, with Edward Walter Maunder (1852-1928) around 1920 and the birth of the Jazz Age. At this point in his life he had been retired from the Royal Observatory at Greenwich England for about fifteen years, having been brought back to the observatory in the middle of World War I. Returned in World War 1 due to a lack of able bodied men, he met and worked with the young Sydney Chapman, placed there as a kind of internal political refugee, since Chapman was a conscientious objector. It was here at the Royal Observatory at this fateful, critical time that Chapman, a high honors "tripos" mathematician (wrangler) from Cambridge delved into the accumulated solar data work of "Walter" Maunder and his better-educated second wife Annie Maunder (1868-1947). Her own rank at Girton College, Cambridge in mathematics was the same as Chapman's—hers, unsurprisingly enough, unrecognized at the time.[47] It was "Annie" who most likely translated a lot of Maunder's data collection into lean mathematical terms for the bright, young Chapman.

Maunder had been at the Royal Observatory full time since 1872, retiring in 1903. He was a pioneer solar astrophotographer and quite modern, methodical collector—and interpreter—of solar science data and it was he who recognized the validity of observations made by Germans like Heinrich Schwabe, Gustav Spörer, and others regarding solar cyclicity, and who tied them together. Notably, it was Henrich Schwabe—technically an amateur—who detected and tabulated the presence of "regular" solar minima and maxima on the Sun in c. eleven year stints. This was the first hint, practically ever, that the sun had any sort of regular or repeating complex behavior at all and as such, represents a stunning achievement in science. Only later does a discovery like Schwabe's coincide with a larger, regular behavior of stars in the context mapped and pointed out by the likes of Jacob (Jacobus) Kapteyn, for example. Of course the regularity noted by Schwabe was one of energetic occurrence and Kapteyn's, regularity of motion, though both these can be linked in ways.

---

[47]   In some instances she is credited with being an Irish astronomer, having won Senior Optime rank (second and third top mathematicians) at Cambridge University, but having qualified for Cambridge in primary and secondary schools in her native Ireland. See for example Bruck, M.T., *Women in Early British and Irish Astronomy: Stars and Satellites*, and Bruck, M.T., *Agnes Mary Clerke and the Rise of Astrophysics*.

(Left to right) (Samuel) Heinrich Schwabe (1789-1875) German (technically amateur) astronomer who discovered a "ten year" sunspot cycle through a lifetime of regular sunspot observation. Rudolf Wolf (1816-1893) Swiss astronomer who quantified Schwabe cycles (11.11 years) for which an extended minimum gap (c. A.D. 1280 to A.D. 1340) is now named for and a superannuated sunspot counting technique (Zurich-Wolf Number). (Friederich Wilhelm) Gustav Spörer (1822-1895) who correctly noted sunspot movement patterns and peculiarities, and who also has an extended minimum gap named after him: the Spörer Minimum (c. A.D. 1460 to A.D. 1550).

Some other persons are worthy of mention as to how things got started in seeing a magnetic connection between sunspot cycles and Earth magnetism, which lies at the base of extended minima and maxima. One was German mathematician Carl Friedrich Gauss (1777-1855) who devised methods of measuring the horizontal intensity of the Earth's magnetic field. There was the observation-collecting English general, Edward Sabine (1788-1883) who founded and made observations from magnetic observatories or "watch stations," and recorded geomagnetic activity throughout the then-extant British Empire. He reported diligently to the Royal Society as well as to the Royal Navy on these matters due to contributions aiding practical navigation. In fact, as early as 1852, Sabine announced to the Royal Society that after systematically measuring magnetic disturbances on Earth from two watch stations in the British Empire (such as Toronto and Hobarton) from 1846-1848, he saw that the geomagnetic disturbances:

Correspond(s) precisely both in period and epoch with the variation in the frequency and magnitude of the solar spots recently announced by M Schwabe as the result of his systematic and long continued observations.[48]

Sabine may have obtained Schwabe's tables, polished and published, by the popular German naturalist Alexander von Humboldt in 1851.[49] This delivered the total import of Schwabe's work to a wide audience, as Humboldt was one of the most serious, high-profile, and well-respected explorer-scientists of his time. Other individuals with respect to delineating extended solar

---

[48]   Sabine, E., "On periodical laws discoverable in the mean effects of the larger magnetic disturbances," No. II, *Philosophical transactions of the Royal Society of London*, Vol. 142, May 6, 1852, P. 103

[49]   Ibid North, p. 466

minima then were (the technically amateur) Frederick Wilhelm (William) Herschel (1738-1822) and ex-professional Richard Christopher Carrington (1826-1875). We note here Herschel's pointing out low growth periods for corn [50] in times of low sunspots, seeing the trend in stock market reports and otherwise overlook this fantastically productive and well-known astronomer. [51] Carrington, to whom one of the so-called laws of sunspot motion is co-credited with Spörer should be tied to Sabine's and Gauss' work, and his fame in this context really hinges on his now-famous observation—"in the act"—of the first confirmed scientific witnessing of a gigantic solar flare—a white flash or even a CME—in his technically amateur solar observatory on September 1, 1859. The magnetic effects of these were recorded independently of him at the magnetic observatory at Kew Gardens by other observers [52] 24 and subsequently 48 hours later.[53]

But by 1920 Maunder was already past the few autumnal years he was asked to return to Greenwich, there for the brief commingling of ideas and observations with Chapman. Maunder had already made his point about the Sun's non-linearity to the Royal Astronomical Society, to include corrected math, and had won the debate versus Kelvin. His best years were well past him and he had in fact but seven more years to live. He was still active in astronomy, having been a founding member of the British Astronomical Association (or B.A.A) some years before. He had taken part in world-wide expeditions for observing, drawing, and photographing the Sun, making significant original observations as well as reconstructing others' work.

Maunder's public hard scientific visibility was extremely low across his career, even for a published astronomy popularizer and well-appreciated lecturer on the subject. He even had a professional job in the field. For the most part we could easily come to accept the idea that Maunder was ignored. To be charitable, it could be said that, some of what he was talking about relative to data and observations of the Sun was not particularly well understood if in some cases, understood at all, in his time. This could very well have been the case. There is no rule holding that all good science knowledge must be well understood at the time of its finding even if people in the audience politely applaud. Rather, if it is valid even if lost, it is rediscovered and subsequently used. There is abundant proof that most discoveries occur long before they are put to practical use. That Maunder credited Germans openly in the latter parts of the 19th Century as regards a useful and correct knowledge of the Sun could have compounded his unpopularity or even dislike in the Anglo-American scientific community of the time. Viewing the lead up to World War I today as a particularly hostile time regarding Anglo-Franco-Teutonic

---

[50]   Wheat

[51]   That is, no sunspots, less Earth food growth. Herschel also had provocative theories of solar structure.

[52]   Due perhaps to Gauss' work, "magnetic" observatories became commonplace and sites of investigations into Solar Systemic-to-Earth connections with magnetism at astronomical observatories and elsewhere in the mid to late 1800s (for example, in Sweden).

[53]   A book treating on the impact of this is *The Sun Kings: The Unexpected Tragedy of Richard Carrington and the Tale of How Modern Astronomy Began* by Stuart Clark.

political and economic relations, it is today easy to underscore great effort by several German scientists as opposed to just two English ones—one actually a soldier. But in those days of rabid colonial aggression in which even the United States conducted colonial wars, tempers ran raw at all levels. Reason and logic suffered terribly in nationalist England and United States. It hardly faired well in nationalistic Germany.

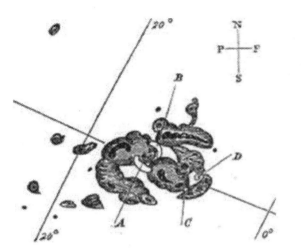 

(Left) What Carrington briefly saw around noon, September 1, 1859, right after making his midday observations. He was on his way out of the observatory when a series of "flashes" struck the white screen he viewed sunspot groups on through his 12 inch scope, only catching the end of the phenomenon. (The sunspot group drawn here is bigger than Earth's diameter.) To modern eyes fixed into space, what he saw on paper looks like what was photographed by the Solar and Heliospheric Observatory (SOHO) some years ago (Right). (Description of a Singular Appearance in the Sun on 1 September, 1859 by R.C. Carrington, Monthly Notices of the Royal Astronomical Society, Vol. 20, p.13-15 (Copyright, RAS) (more likely, Balfour Stewart, MRNAS, 1861)

Maunder foolishly or bravely took unpopular personal stands,[54] spoke out about them, was a devout Methodist and had strong scruples. He was, in fact, even an exacting religious scholar.

But England had an Empire: Germany wanted part of that and a bigger one. There was Maunder, a Crown employee (and a low one) actually calling the "laws of sunspot motion" "Spörer's Law," and arguing the validity of Spörer's point to Englishmen who would or could not swallow it. He credited Schwabe in print and praised him for his work. Later on, while the chaps were being machine gunned out on the front, there was Maunder, having one on ones with a conscientious

---

[54]   Such as getting his wife into the Royal Astronomical Society, for example, and making comments about universal suffrage for others in her position: his even-handed scholarship on other religions, etc.

objector, even if Chapman had been a wrangler. [55] Maunder's aim was simply and purely the recognition of scientific validity regarding the Sun's functioning. (It is doubtful he had shares in African/Chinese-German financial ventures for example.) Thus sat Maunder, still corresponding in astronomy into the 1920s.

◊◊

The oddly-derived and difficulty-linked detective work into seeing how the sun could weaken or seem to shut down magnetically came from the science of botany and related tree studies, of all places. But in this very rare case, the "botanist" (actually a tree scientist) had also been an astronomer. Andrew Ellicott Douglass (1867-1962). Vermonter by birth, Harvard man and observatory-scout for wealthy dilettante astronomy popularizers, he was about mid-way through his second career as an astronomy professor when he came across something in tree samples, which he had an idea for dating solar activity with. He looked up a man in England—Maunder—who supposedly had amassed as much sun data as *he* had collected tree sample data. A brief flurry of correspondence occurred.

Douglass earned his bachelor's of science in astronomy from Harvard University and after being fired from Lowell Observatory,[56] taught at the nascent University of Arizona. Unusual for scientists of any generation he made a career switch to archaeology while still being an astronomer (which would have made him a proto "archaeo-astronomer" as well in context of his ancient tree ring-sun connection). He became "the father of" dendrochronology as that old way of crediting original scientists went. Dendrochronology—other than being a mouthful to pronounce—is dating things from tree rings that took on more panache after the discovery of Carbon-14, unknown at the time of Douglass' pioneering studies in tree-ring dating, however. The scientific insight revealed here is uncanny. His converging botanical observations with solar science's repetitive cycles is nearly profound. While analyzing several species of trees and their rings' widths and other anomalies he made an observation, while assigning dates, as to what he believed to be the near absence of c. nine—c. 13 year solar cycles (Schwabe Cycles) in the rings over certain ancient time periods, regardless of the species. As an astronomer, he was aware of what Schwabe Cycles were and most likely, details on their numerical peculiarities.

---

[55]   No getting around this part of his manliness—even if there were women who were mathematically his equal or near-equal. Chapman's non-militant streak is shown in his work with the German, J. Bartels, in the run-up to World War II resulting in the co-authored book we owe a great deal to as regards the Sun-Earth relation, mathematically: *Geomagnetism* (1940). In this, he obviously develops ideas from observations handed over by the Maunders. Chapman spent his later career at the High Altitude Observatory in Colorado, USA.

[56]   Douglass had been, in effect, Lowell's professional astronomer. He was replaced by Vesto Slipher. Douglass had located Lowell's observatory for him, and prior to that, had located a South-American based one for Harvard University.

"E." (for Edward) Walter Maunder (RAS Library) early in his career.

Many of his European tree ring specimens for the Seventeenth and Eighteenth Centuries had very thick rings for many years, meaning much moisture (which in turn meant <u>much</u> cloudy weather and much precipitation = cool). But the signal for missing "11-year cycles" in the tree rings was clear: especially in the record after 1620 until the mid 1700s, which Douglass, upon comparing the data, claimed that the "flattening" thereof was "striking."

In a letter to Maunder,[57] Douglass said he had been studying yellow pines regarding these 11 year "absences" across the tree ages dating from the late 1400s, the 1500s, 1600s and 1700s and noticed that just after what is now called the Wolf Minimum, there was a pronounced "absence" of these in what was later noted as the Spörer Minimum, a bit more of a lack of them, and then, after 1620 to c. 1680 (the next sixty years after 1620) the "flattening" (vanishing) of the "curve" for 11 year cycles became very noticeable and pronounced, and that Sequoia trees in particular revealed this flattening, or near disappearance, of 11 year cycles. (That Schwabe Cycles vanish altogether at times is theorized.)

Douglass before one of his specimens.

---

[57]   Letter Maunder read from A.E. Douglass before the British Astronomical Association in April, 1922.

Some of these trees (Sequoias for instance) could obtain great age indeed: a few thousand years. Since Sequoias live in some cases over 3,000 years it becomes quite a "computer-printout" of solar activity regarding its tree rings' accumulation (or non-accumulation) of such into deep time. Douglass remarked in the same letter to Maunder that the Sequoia samples were strongly flattened, recording no solar cycles at all it would seem, all the way to 1727. Douglass concluded, as essentially one professional astronomer to another, that since about 1400 all the way to the end of the Seventeenth Century the sunspot cycles had been operating with "interferences." What these interferences were no one could tell. (It is still anomalous.)

Here, then, was a data-detailed terrestrial fingerprint for what Maunder had been seeing and studying for years, and which he must have long suspected, with comments from a fellow astronomer, this one from Harvard. It was from a source as far afield as still-Apache dominated Arizona, a US state for only ten years in 1922 by a professional astronomer he did not know personally, and who was carving out the field of dating solar oddities by using tree rings—something the scientist-never-credited-as-being-a-scientist, Leonardo Da Vinci, had thought possible several centuries ago.

Maunder had had an earlier hint of an Earth fingerprint of the Sun's behavior upon it. Twenty-two years into a hectic and overly-committed career that frankly wore him out, in 1894, Maunder had written on prolonged sunspot minima [58] and his wife's Annie's close friend and fellow Irish woman scientist [59] Agnes Clerke then made some very cogent observations on his article that Maunder no doubt concatenated with Douglass' data later on.

---

[58]   Maunder, E.W., "A Prolonged Sunspot Minimum," *Knowledge*, August, 1894

[59]   Like Annie Maunder an observational astronomer who eschewed the role offered to her at Greenwich. She became an independent historical scientist instead, concentrating on astronomy. She was perhaps the most respected astronomical history author in her day, and her analytical points, concatenated, as shown here, are not to be snubbed.

(Left to Right) Leonardo Da Vinci (1452-1519). Agnes Mary Clerke (1842-1907)

In effect Clerke—then one of the most well-regarded astronomy science-writer/scientists—drew together the reduced number of sunspots over longer periods Maunder noted in his article, to a probably simultaneously-occurring "magnetic calm." Gleaning from eyewitness reports in the 1500s and 1600s amounts of auroral displays prevalent in England at those times, there had been, she reported,[60] an absence of auroral sightings throughout the 1600s, compared to the mid-to-late 1500s. She hazarded that lacks of auroral displays were tied to magnetism, probably of solar origin. Then she tied these to the lack of sunspots. In her response to Maunder, she carefully couched her words.[61]

By 1922, then, Maunder saw from data "snapshots" noted as having taken place in the late Fifteenth through mid-Eighteenth Centuries at least three things occurring simultaneously:

- Lacks of sunspots (maybe all, at times?) compared to other times
- Lack of auroral displays (at times—hence, magnetism dissipates)—Clerke's brilliant and insightful linchpin point (but not to be confirmed for years)
- Lack of 11 year Schwabe Cycles (the apparent disappearance of solar maxima, especially at some times more than others) as their absence in tree rings tells, regardless of species—Douglass' quantified data report revealing this

---

[60]   Clerke's analysis and comments on Maunder's August, 1894 article in *Knowledge*, September, 1894

[61]   Actually aurorae *were* visible in England then: but they were, *indeed, extremely few* between 1620-1700. Aurora were much more common after 1715, in England, and in the Northern Hemisphere generally, diary etc. records have shown. Research has vindicated her cogent conjectures regarding the connection between reduced magnetism and low sunspots.

This is the recondite scientific summary of the Sun and Earth connection as Maunder could have seen it roughly by 1922, in diagram form:

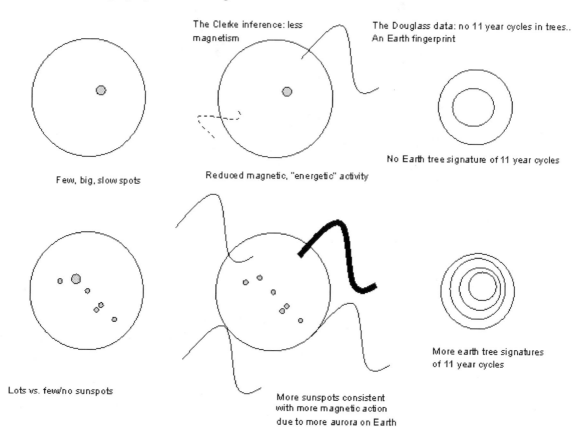

The Clerke inference: less magnetism

The Douglass data: no 11 year cycles in trees.. An Earth fingerprint

No Earth tree signature of 11 year cycles

Few, big, slow spots

Reduced magnetic, "energetic" activity

Lots vs. few/no sunspots

More sunspots consistent with more magnetic action due to more aurora on Earth

More earth tree signatures of 11 year cycles

Maunder also realised that the sun's inner action was somehow different from the sun's surface action. He was also aware of a co-rotational aspect to energetic motion off or on the Sun as the Sun, itself, rotated.

This was all before 1930. It was very strange then. But it is still strange. Maunder died in 1928.

William Thomson (1824-1907), later Lord Kelvin, as a young man.

By 1930 his work starts to get buried for the next 40 years, until the next courageous thinker on such ideas takes it up. His story follows.

◊◊◊

It is not too odd that sailors are drawn to studying the Sun for obvious reasons related to the antique art of locating themselves in the world and the day with the Sun and a sextant, among other things. There was (actually soldier) General Sir Edward Sabine after all, and John H. Lefroy, another soldier and Sabine's collaborator on things solar and terrestrially-magnetic. But they used the navy to get around quite a bit from magnetic-observatory to magnetic-observatory (as far afield as St. Helena and Toronto) in their day. Their most practical aim was to collect data in assisting sea navigation. Another thing to keep in mind about soldiers and sailors is that out of a sense of duty they take risks.

"Jack" (John A.) Eddy (1931-2009) (US Naval Academy)

About the time the Space Age was being born in Russia, young Lt. Commander John A. Eddy was just getting out of the navy and entering graduate school. Eddy got on to a program at the University of Colorado in a course of studies combining solar and geo physics. And so began a tortuous career.

Due to either the foolishness or the bravery of some the rest of us generally benefit. Being an omnivorous and curious investigator and fearless of the bureaucratic brass Eddy's work wandered over different disciplines, and he faced the usual barbs for it (in this case, at the High Altitude Observatory—HAO). Eddy went off topic in his studies. When they tell you to study just the peas on the plate and let someone else do the carrots and potatoes and meat they mean it. Budgetary cut backs in the 1970s hastened Eddy's dismissal at the HAO.

Ruminating upon astronomical and geophysical subjects both various and profound relative to the Sun and down to Earth's climate, Jack Eddy took up the matter of the previous several pages in this chapter and basically revived and modernized Maunder's work. He did this, essentially career-less, by sliding his nameplate over a door in a room at the Harvard-Smithsonian Center for Astrophysics, in Cambridge, Massachusetts, supported somehow by the labyrinthine American NASA-Smithsonian Institution-National Science Foundation nexus. As a temporary officer of the Smithsonian he had access to some of greatest untouched-by-bombs-or-ransacked-by-wars libraries in the world. Whatever he asked for, librarians delivered it to him in government envelopes. Forget the Harvard part, and what you have is the Smithsonian Institution; the perpetuation of which was guaranteed by John Quincy Adams for millionaire Englishman Smithson's wish that America (a country that had been very good to him) have a museum like Britain's. The money for research done by the likes of Eddy is supplied by that trust for the most, less by the government tax dollar and Harvard University. It is why you can enter the museum in Washington, D.C. free of charge. It is the property of the people. The Smithsonian is where the buck stops in many matters where career interests or beliefs mix and halt others.

By Eddy's time a knowledge and use for isotopic time-stamping had come to light. So he was taking his working knowledge and connecting it with the isotope used for seeing into deep time as described in detail in Chapter 1: Carbon-14. This was perhaps the first of the many "proxies" (proxy isotopes) used for deep time dating. By the mid-1970s most all of what Eddy applied to the Maunder Minimum (outside of naming it, "the Maunder Minimum"[62]) regarding Carbon-14 and its application to solar cycles was already either suspected, known, or well known by laborers in the salt mines of science.

Telling of what a great counter C14 is without going into its fascinating history is like overly promoting insights into nature as witnessed by witch doctors from Borneo 100 years ago. It makes it sound like all magic without the grist and grit of the great scythe of science winging in the wind in between. Willard Libby, one of the Manhattan Project [63] people in World War II, had the Carbon-14 method down pat as a workable theory early in 1949. This is how it goes, roughly speaking. Trees (as well as all plants) absorb Carbon Dioxide from the air (which is plants' "air") and in the radiant light / Carbon Dioxide shift they use mostly to make sugars (and

---

[62]   Eddy could have been criticized by certain Germans, saying that the minimum "type" should have been named after Spörer. Maunder himself may have agreed with this.

[63]   That is the inventors of the atomic bomb

exhale Oxygen, the positive feedback of which provides us with our "air" in "photosynthesis") *trees* absorb Carbon-14 (C14) that is near to the amount of the atmospheric isotope of C14. When in this case trees pass away or are otherwise used (cut for lumber, are burned, etc.) the tiny amount of C14 drops at a very particular rate, since the C14 is radioactively decaying. It is this tiny bit of stuff then that allows the total sample to be aged. To get the C14 "date," samples (of the tree, say) are arranged, a ratio is obtained, and a calibration sheet is used to read off the sample's age from the ratio.

They test for it like this. First, samples are soaked in Sodium Chloride followed by a Sodium Hydroxide bath to clean off the gunk, sometimes taking away valuable evidence; but, such is how it is in the land of proxy-dating—which is why so many can and will dispute its relevance. The isotope's structure is not touched, although this is debatable and so, clouds more doubt around its usefulness. A selected sample portion is then control burned.[64] Again, some data can be altered or ruined. Last, the age is read from a chart; the age versus the ratio. This isotope is free-floating in our atmosphere, where it is hammered by cosmic rays up in the ionosphere, and the Carbon-14 deep-time dating method is good for anything living up to 60,000 years ago (after that it is up to other isotopes, like those of Uranium-235, to use for dating fossils etc.). The "bias" in "Carbon dating" can be wide and high: up to and over 5,000 years from whenever the stuff living collected the isotope in its biomass (a bone or in the case of Douglass' discovery, a piece of tree wood).

But the "breakthrough of 1949" by the great Libby was not enough. By 1951 the search for the science of Carbon-14 dating was bogging down for all its early promise. The gaggles of geologists, archaeologists and (fewer) paleobiologists, geophysicists and astrophysicists waiting for this new scientific tool lined up to see what the mathematical physicists and chemists had wrought from raw Nature. Much research monies had been raised to lift the hopes of all this in Britain, under Harry Godwin, Alfred Maddock and others. The result was getting a hungry graduate student to put it all together. Paraphrasing the words [65] of a "third man," "the genius rank" had established the path (Libby). Now, it was up to the master workmen to flesh out the details. What would be the counting gas of the radiocarbon? "Gas proportional counting" was the "way to go." But it didn't occur to any of the Americans and Englishmen to use Carbon Dioxide. This came from Hessel de Vries (1916-1959) a Dutch physicist at the University of Groningen. The British combed the literature and found it in a letter to *Physica* [66] compiled

---

[64]   Mass spectrometry technology (Accelerator Mass Spectrometry, or AMS) has since allowed for extremely small parts of objects to be burned, saving in some cases, precious samples. A good treatment on hard cases that are very contentious and debatable in archaeology is found in *Quest for the Origins of the First Americans* by E. James Dixon (University of New Mexico Press: 1992).

[65]   Much of the Libby-to-de Vries connections here come from a single source: *Radiocarbon dating in Cambridge: some personal recollections*: A Worm's Eye View of the Early Days, by E. H. Willis. History of Quaternary Research in Cambridge webpage.

[66]   XIX, p.987, 1953, according to Willis

by de Vries and G.W. Barentsen [67] in 1953. The letter was titled, "Radiocarbon Dating by a Proportional Counter filled with Carbon Dioxide." When the "third man" visited De Vries in the Netherlands to hire him, this is what he confirmed:

The answer to the carbon dioxide riddle was simplicity itself—it was extremely sensitive to the presence of electronegative impurities, such as sulphur dioxide, and you had to purify the gas to better than one part in ten million, and only then did you obtain decent counting characteristics. The voltage required was very high since the gas was going to be used at three atmospheres pressure to get as much carbon into the counter as possible—but at last it was proven that it could be done! It was interesting that even in 1955 Bill Libby confided to me in a corridor that he doubted it would work! [68]

So even the "genius rank" doubted it: according to the humble third man. He and Libby took a few drinks as they watched the Sun go down over Santa Monica where Libby lived as they hacked out the idea of radiocarbon dating. Note well that though Libby doubted, he did not deny.

Hessel de Vries: the first to see the physics of the Sun-earth climate connection via "proxy" isotope data.
(Instituut voor Nederlandse Geschiedenis)

De Vries had done other things, like discover insects' ability to detect the polarization of light in the sky,[69] which shed light on applied entomology in agriculture for the purposes of knowing what bees do, for instance, and how dependent they are upon Sol to detect pollination targets and even be active. Their sensitivities to Infra Red (IR) was pointed out. A few years later, in that apparently pivotal year 1958, de Vries: showed that there were systematic anomalies in the carbon-14 dates of tree rings (Douglass had seen lack of Schwabe Cycle evidence here

---

[67]   Another student was Minze Stuiver, also a figure in the Sun-earth climate connection story.

[68]   Ibid, Willis

[69]   *Optics of the Insect Eye* (New York Academy of Sciences, Annals) along with Kuiper-belt discoverer Jan Kuiper.

from the same steps of this ladder much earlier [70]). His explanation was that the concentration of carbon-14 in the atmosphere had varied over time by up to 1%. He hypothesized that the variation might be explained by (a) something connected with climate, (b) that it was not created in the atmosphere at a uniform rate due to variations in the Earth's magnetic field, or (c) *a cause lay in the Sun itself.* [71]

De Vries "gave enthusiastic and continuing support to the Cambridge Laboratory until we were all saddened by his untimely death in the early sixties." [72] Well, actually, like Ritchie Valens, Buddy Holly and the "Big Bopper," he didn't make it to 1960.

This is where the story of the climate and Sun and tree rings gets tangled up in the branches (excuse the pun). It was the loss of a brilliant man. But the study of the Sun-earth connection is chock-filled with these sorts of tales.

The kind of analysis done by De Vries about the tree rings, the variation of Carbon-14 in the atmosphere, and putative causes laying in the Sun were very "left field" at the time. Most matters of space revolved around a political "race" between the bilateral symmetry of a post World War II world (the USA and Russia) and with all the little players in between. There were some geologists interested in C14 dating. If even heard of then elsewhere, it must have been taken as a curiosity in the annals of the strategic weapons making network, all concentrating money and brains on firepower and numerical leverage on atomic and related weapons worldwide, and then the space race. Few were looking up at the Sun in this regard so to speak. Whether Nature had anything at all to do with our ultimate well-being went between the usual journalistic Scylla of death by ice and Charybdis of death by fire. In 1959 Betty Friedan, later known for the politics of women inadvertently chimed in early on the politics of climate, writing in Harper's an article called "The Coming Ice Age; A True, Scientific Detective Story." [73] As the bomb designers rolled out their ever-more-advanced ideas/products to the military industrial complex, the accompanying scare journalism went on deeper in this bent, with threats of "nuclear winter" after a major hit-the-mark swap of superpower mega tonnage in bombs that would alter the Earth's climate in such a way as to induce massive climate change for the cooler. As a staff editorial department worker at the Stockholm Peace Research Institute (SIPRI) near the end of the Cold War this writer [74] edited a wealth of such literature, in essay, article, and

---

[70]   Author's insertion

[71]   The sources for this quote, are: Willis, E.H. (1996), *Radiocarbon dating in Cambridge: some personal recollections.* A Worm's Eye View of the Early Days, *Vries, Hessel de* (1916-1959), by J. J. M. Engels, and *The Discovery of Global Warming*, by Spencer Weart. (Italicized portions by the author.)

[72]   Ibid, Willis

[73]   So well respected was the level of English used in this essay it became an example of how to write an essay in a high school English text, especially a persuasive science essay. *Adventures in American Literature* (Laureate Edition) Edmund Fuller and B. Jo Kinnick (Harcourt, Brace & World, 1963).

[74]   This author

even book form. "Death by warming" did not really take hold until the late-1980s, with United Nations backing and support.

But in 1976, former Lt. Commander Eddy, USN, and PhD; erstwhile investigator at the HAO, and now on the ropes at the Smithsonian, searched for a new career. In the process of doing so he started investigating solar minima and maxima with isotoptic evidence as its tracer. The following is an amalgam of his work, though little of it is drawn from the source work that makes his "Maunder Minimum" point.[75] The outline, given the C14 isotope measuring and its description above, connected to the observational and theoretical groundwork laid down by the Maunders, Clerke, Douglass, and all others discussed so far is first seen in this graph in Figure 2. Given all we know now, the squiggles and bumps, spikes, valleys, and numerical notation all start to make some sense.

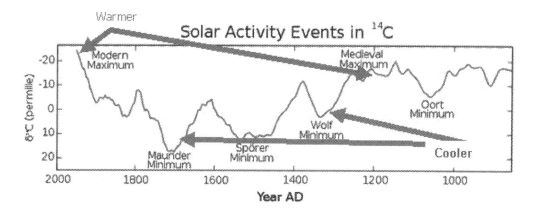

Figure 2. Radiocarbon "report" of solar cycles and C14, along the lines of isotope "testing" and the results from tree rings-to-solar-cycles—showing the strangeness de Vries postulated in the late 1950s. This ties in Douglass' work, the Maunders', and Clerke's. This work was by Eddy in the 1970s.

In Figure 2 the C14 isotope "traces" the apparent global cooling (it is not causing it) by its appearance as an isotope, as well as the global warming (C14 also is not causing it) by its absence. On the graph's left side, upward and downward on a coordinate plane, there is the amount of C14 (delta, or "Δ") per mille. Along the bottom, you have the years counted out by tree rings and invariably, the Schwabe Cycles: A.D. 1000 at the far right, to A.D. 2000 (or, about now) to the extreme left.

That the Medieval Maximum and putative Modern Maximum recorded high negative (or fewer) amounts of C14 in tree rings at their peaks is no surprise, since Sol, strong, blew much of them past our magnetosphere. That the valleys seen in the Maunder, Spörer, and Wolf Minimums

---

75   Eddy, J.A., "The Maunder Minimum," *Science*, 18 June 1976: Vol. 192. no. 4245, pp 1189-1202.

show high positive (or, more) amounts of C14 in the absorptive rings is also no surprise. This was since Sol, very sleepy, let them blow out, but, consequently Earth's magnetosphere, being very relaxed due to no solar wind flattening it back, allowed a lot more cosmic rays (higher rate of C14 production) into the Earth's atmospheric envelope. From here it floated on down, wandering, being absorbed by, among other things, trees, in that gradual way previously mentioned.

Thus do they conclude that it must have been cooler in the Spörer, Wolf, and Oort Minimums (even if we don't know much about what Earth physically looked like at that time, from records, pictures, etc.). This not due to that isotope of Carbon, C14, but since the abundant finding it at the Wolf, Maunder, etc., periods meant much meandered into the atmosphere due to a magnetically—to quote Clerke—inactive Sun. We rudely conclude that it must have been warmer in the Medieval Maximum due to the obverse pattern. That is, that a magnetically active Sun blew the particles—C14 being just one—out of the atmosphere, since very little C14 is found in biomass dated to the Medieval Maximum. A very active Sun must mean, then, a warmer Earth, somewhere, and at X temperature (delta) though this latter parameter might never be quantified. The opposite is, of course, that the Sun, much less active, must mean that Earth is cooler somewhere and at X temperature.

More concrete cultural glimpses are of course, detectable, giving an idea that it was warmer then, and these can be qualitatively superimposed over such data as these to create a basic picture, as will be done in Chapter 3. As regards the Medieval Maximum, for instance, there is some cultural information that lends weight to the warming across the Northern Hemisphere. Scandinavians (Vikings, then, to a man and woman: Norwegians and Swedes and Danes did not exist yet) saw fit to settle the Faeroes, Iceland, Greenland, and even parts of Nova Scotia from c. A.D. 900 onward till about A.D. 1450. Their descriptions in old Norse, translated into modern languages, tells us of verdure and abundance of game, and the ability to farm and to reap; to foster and cull. They tell us why they called Greenland "Greenland," for example. Apparently around A.D. 950 greenery was quite abundant around the icecaps. The explorers' boats were mostly open to the elements—saying something of the physical strength, stamina, and tolerance to cold of these pre-Black Plague Nordic peoples, perhaps more Neanderthal than Cro Magnon. It also, however, says much about the air temperature over the North Atlantic Sea of the time.

Then at the opposite extreme there was the Maunder Minimum, the dates of which coincide with the first raw fruits of modern observational and recorded science. A lot of biographical and descriptive data of the goings-on in the climate at the time exist, and being the lowest lower "bump" in the C14 collected by Eddy, tells us a tale that may have been the less so, were it only so deep as the Oort Minimum.

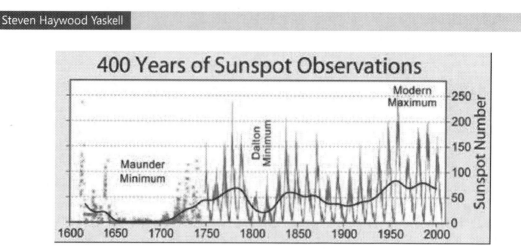

Figure 3, Sunspot observations (reconstructed from c. 1600 to c. 1860) and sunspot number.

Figure 3 regarding sunspots can be interpreted together with Figure 2, the one about C14 levels. If you look at them back and forth a few times it starts to be seen. Sunspots mean a more *active* sun, if you accept the premise of isotope proxy data of the C14 type. The sunspot *number* (counts of how many were seen and recorded on the Sun's face) is in the coordinate plane in Figure 3 to the far right, the Earth time heading more logically perhaps, from left to right on the coordinate below (A.D. 1600 to A.D. 2000). The black line running through it is probably either recorded or reconstructed "Total Solar Irradiance" ("TSI") and note that it follows, very roughly, the peaks and valleys in the sunspot number counts.

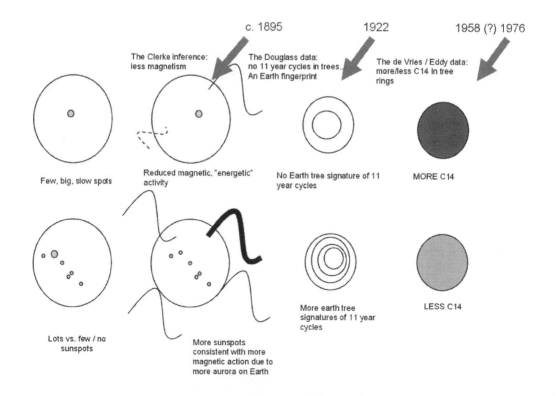

The above chart is the still-recondite scientific summary of the Sun and Earth connection as Maunder would have seen it roughly by 1977, had he been alive, and no other progress having been made till that time in these matters except for Eddy. Maunder would have seen from the evidence that a weaker Sun lets more particles be absorbed into Earth past its own magnetic sheath. He would probably have caught on to this quite well and cogently remarked upon it.

Though they thought that solar cycles disappeared, it was later revealed that the solar cycles (Schwabe, etc.) did not vanish altogether. What happens, at times, is like what was shown in the start of this chapter in the amplitude chart. If you take a look at the C14 graph above, and then the diagram (below) you will see that the graph—deep in the Maunder Minimum "valley"—had little maximums. And the Medieval Maximum "peak" had little minimums.

It is just a difference in scale, or perhaps magnitude, would be a better word choice. Magnitudes are, of course, signs of what old high school science texts would refer to as force vectors. Magnitudes can be assigned to the severities of earthquakes regarding force. They are also assigned to the apparent and absolute values in the brightnesses of stars and behind this brightness could very well be the force of magnetism. That is, literally, a magnetic force field.

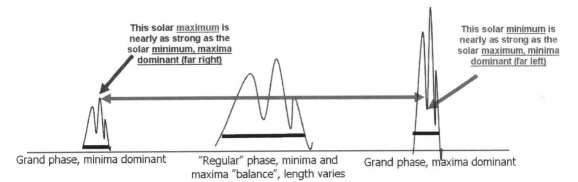

This solar maximum is nearly as strong as the solar minimum, maxima dominant (far right)

This solar minimum is nearly as strong as the solar maximum, minima dominant (far left)

Grand phase, minima dominant   "Regular" phase, minima and   Grand phase, maxima dominant
                                       maxima "balance", length varies

Sometimes the ups and downs are sharp but low—a GRAND minimum. And sometime they are sharp but high—a GRAND maximum . . . and their gap widths as occurrences are possibly equal in width.

Stellar activity never stops until, of course, it dies out altogether with the star itself. Even then some stars produce energy but in altered form. Our sun is bound by the same natural law so far as it is known.

Life has existed on Earth now for c. three billion years and will end on it when this extraordinarily deep future point in time is reached. As more complex forms evolved from protolife, perhaps to include Man, so did our Earth's atmosphere likewise evolve in chemical construction, complexity, and energy transfer complexity most likely. Earth's atmosphere evolved in step with the life that lived within its cowl and its life probably helped make that very same atmosphere in part.

How does life on Earth fare in the meantime? Especially human life? And what of these solar extended minima and maxima effects on Earth's population over longer time periods—and even over fairly recent ones? The evidence to be shown forthwith is mostly of a "smoking-gun" sort; it is qualitative, and is just now being partly quantified.

We turn to this subject in the next chapter.

3. Grand solar phases in possible civilization-altering contexts

Has our Sun been responsible in moving civilizations through time? No deterministic approach is made here nor any religious or quasi-religious one. Side-stepping national socialism and scholasticism we enter into the ancient halls of the end of the deep ice age of 11,700 before the present time (B.P.) to c. 9,000 years ago. Then a closer look is taken from about 1,600 years ago to the present for a cursory view of trends in human civilization. Overlaid globally on this is the Sun's insolation—that is, the total exposure of its rays—over the periods covered. Significant to this discussion, the Hallstatt Cycle (or oscillation) of 2,300 years and alternately 2,250 [76] years is introduced, particularly for later consideration in long-term solar-induced climate change. It is placed here due to its context with the aforementioned 1,600 years of human history that shall be telescoped sharply into view.

Solar forcing is small when compared with what it does to Earth over time, the longer periods—the millennia—giving experts a clearer picture than the centennial or decadal. The pictures presented by these latter offer little and can hardly be grasped. We hazard in turn a look at data for 1,600 years against human civilization rises and falls (for want of a better phrase [77]) and what may cause these (plague, drought, war) etc. The visual approaches used are simple (filtering via sliding window) although this is weak compared to advanced wavelet transform analysis or the Fourier power spectra analysis when applied to something like Earth climate.[78] A simple graphical comparison of filtered data is fast and provides a detailed "look" at data, thereby reducing the coloring and destruction of the information that often results from statistical methods. (This is not intended to negatively color statistics in studies.) There are in addition the very often poorly understood limitations and constraints of statistical tools commonly used in constructing such views.[79] In another case, however, a link based on statistics between very recently-recorded conflict and climate *is* shown.[80]

---

[76] Cliverd (2,300) and Steinhilber (2,250) as cited in de Jager, C., and Duhau, S., "Sudden Transitions and Grand Variations in the Solar Dynamo, Past and Future," *Space Weather and Space Climate* (in press, 2012)

[77] My apologies to Edward Gibbon, author of *The Rise and Fall of the Roman Empire*.

[78] Bill Howell (private communication)

[79] This entire paragraph is gratis Bill Howell (2011).

[80] Hsiang, S. (2010)

The Holocene in a cyclic view of climate events

What kind of world was it right before—and right at the end of—the last deep ice age 11,700 years ago? Why take such a stepped-back view from the Holocene, which is a little more than ten thousand years in duration? This view is long in one sense and short in another.

Mainly this is done to lend perspective to graphs that shall show the total averaged solar insolation construed as TSI which is then superimposed over human civilization for the last 9,000 years. This view lends greater breadth to deep time, visually, up to the present. To help do this, the isotopes Oxygen 16 and 18 (^{16}O and ^{18}O) are introduced here to lend greater weight to understanding natural temperature changes in the past.

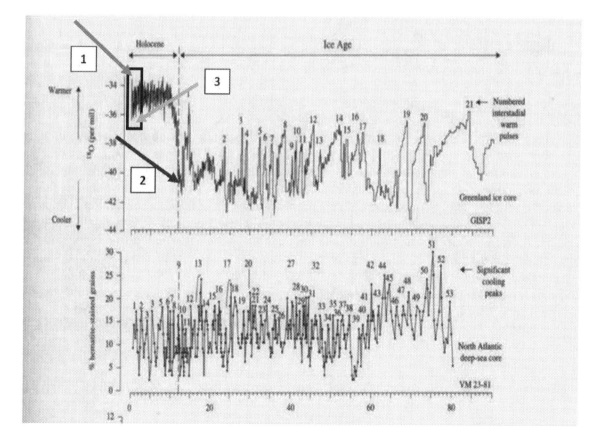

Figure 1. Cyclic (or pseudo) climate-event dating: ice rafting" events" from ocean sediment core samples, showing positive and negative relationships between the ^{18}O and ^{16}O isotope. Arrow 1: current upward trend in warming, arrow 3: cool trend ending c. 1880, arrow 2: far colder world c. 11,700 years ago (After Gerard Bond, 1999)

Figure 1 is based on Gerard Bond's research on rock or stone (hematite) stains, moving glaciers, and a tie to solar activity. It covers c. 90,000 years for at least four full Earth precessional [81] periods leading up to the early Holocene, when the Earth was in a deep phase of the ice age. But even in that deep ice age there were at least 25 rapid climate fluctuations peaking the amplitude of climate on Earth upward in relative warming trends. That is, there were warming peaks even in much colder, deep ice age times.

"Up and down" wave amplitudes are shown by small hematite grains (in Figure 1's case) found in sea off the Northeast US and Canadian coasts. The hematite was found like this: the icebergs carried the stone from the land. Ice as such then "rafted" on water. Such stone debris in ocean core samples taken by scientists came from such melted icebergs that broke off the tongues of glaciers. These icebergs contained the small stones (rock grains) that fell to the ocean floor as a consequence of icebergs melting on their way south (becoming part of the sediment and so, collected by researchers as core samples thousands of years later). Collected by taking ocean sediment cores, they were then isotopically dated. Delta counts of the Oxygen-18 isotope in the core samples, compared to the hematite stain grains reveals a c. 1,500 (+/-500) year "hop." These would be, then, the recurring rapid climate fluctuations (25 or so) just mentioned, and these are highly debated as to their cause and prevalence (see the "numbered interstadial warm periods" noted in Figure 1 in the top graph).

The "hop" is the temperature changes at the respective times, recorded by comparing the two proxy isotopes of temperature change. These two are Oxygen-18 and Oxygen-16. Researchers then obtain in such a comparison (like in Carbon-14, for duration) a deep Earth date, from a ratio, off a chart. In the case of Oxygen-18 and Oxygen-16 the time is stamped to the temperature. If the ratio of ^{18}O to ^{16}O found is HIGHER, it was COLDER then. If the ratio of ^{18}O to ^{16}O found is LOWER, it was WARMER then. In Figure 1, these WARM amplitude peaks are plotted upward (the top graph in Figure 1).

In this theory, within a 90,000 year period there seems to have been a warming rise recorded by these two isotopes regardless of whether or not a deep ice age was ongoing. Figure 1 shows at least twenty of these interstadial warm pulses prior to the deep ice age's end 11,700 years B.P. and that a steady rise in temperature hemispherically occurred rather quickly thereafter. Observe the rise in the amplitude from the arrow 2 to arrow 3 in Figure 1. It is a whole new plateau, so to speak, in globally averaged temperatures, as weak as this is in such a context. The Earth's mean temperature must have risen, since the temperature isotope's ratio roughly today is -34 ^{16}O-^{18}O, and at the end of the deep ice age, it was -40 (higher, in other words, and so, colder back then in at least regional or specific locations). Figure 1 reflects all this. The very last of the ice age fauna like Woolly Mammoth were on their way out. Neanderthal, our human cousin and possibly chief architect of survival in the cold of the deep ice age, was long gone. Neanderthal's descendants had by this time long learned to read the shifting climates.

---

[81]   Earth's axial tilt, which gives us new pole-star orientations every c. 20,000 years.

As humans we are hardly any longer the hot-weather type of proto-human that once roamed in tropical periods three million years or more ago in equatorial Africa—when far-off Britain's climate resembled that of the Philippines today.

Arrow 1 in Figure 1 shows the current upward trend in temperature according to these isotopes of oxygen. The lower dip inside of this high rise (arrow 3) shows the end of the last cooler trend about the year 1880. Visually it is a small distance and indeed, in deep time, it is just a short distance as is plainly shown by the spread of all the years in the figure. But near ancestors report down to us from the late 1800s show how noticeably snowier and colder, windier, rainier, and wetter it was a mere 100 to 130 years ago.

Eleven thousand years ago, close precursors of modern humans had been around for more than 40,000 years. Their cousin the Neanderthal, still in existence at Cro Magnon's beginning, may have taught them how to survive the cold. The two may have intermingled. The ice fields and sheets started to recede in the US Northeast (close to the Laurentian if not a part of it) to name one well-known location. It took a few thousand years for it to re-populate with fauna and flora remotely familiar to what we see there today. In fact the US Northeast is still quite hilly, rock-till and boulder decorated characterized by eskers, terminal and lateral moraines, kettle holes and other evidence of moving ice mountains' ravages of the land that is barely covered over with vegetation even now. Predictably as one reaches Canada the moraines, eskers, etc., are more visible and in many cases higher alternately lower/deeper, overall, flat plain areas aside. As shown in the top part of Figure 1, the amplitudes of the Holocene are very small and tightly knitted compared to those in the deep ice age to the right in the figure. Still they show their up and down variations for the cooler and the warmer, us currently being in the warmer. The current "down" part of the ^{16}O-^{18}O ratio (and higher wave peak) means the water and air over us are, roughly speaking, warmer since 1880 in this most recent leg. In the vast scheme of things, that timeframe was cold; but not that cold in the Northern Hemisphere, compared to now.

In Figure 2 (blown up section of Figure 1) we get a glimpse at the last 9,000 years of the Holocene and its tightly knit, up and down peaks of cold and warm periods before 10,000 years ago. The warmest times were far from the coolest ones in the Holocene, a few exceptions aside. The Holocene is indeed a new plateau for Earth climate behavior after 9,000 years B.P.

Around 3,000 years after the deep ice age's end, there was a significant cooling period called the "8.2 Kiloyear event" (Kyr) or, 8,200 years before present (or 6,200 BC). In Figure 2, the downward peaks are such that the 8.2 Kyr event is nearly as far downward as the much-more recent Little Ice Age (LIA). For the 8.2 Kyr event's occurrence we look at oscillations in the ocean current system. An abrupt cold period lasting around 300 years occurred around 8,200 B.P. in the North Atlantic area. In Greenland ice-core records measured by scientists, it is characterized by a reduction in temperature greater than 1° C, a decrease in the ice accumulation rate, increasing

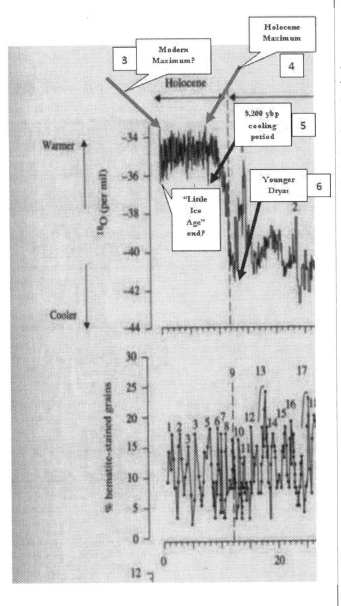

The 8.2 Kyr event, the Holocene Maximum, Henry's Law, and today

Arrow 3 to the far left is today. Arrow 4 to the right covers the time period of roughly 6-7,000 years ago known as the Holocene Maximum. Observe that the "warm amplitude peaks" are about the same as the present data for 18O and 16O ratios of the oxygen temperature isotope, and rather high up with less severe downward millennial "down-peaks" overall.

Arrow 5 points to c. 9,000 year before present and refers to the c. 8,200 year B.P. cooling period. Arrow 6 points at the Younger Dryas (c. 11.5 Kyrs B.P.)

This downward spike could account for the ice rafting events recorded by Bond and the hematite stains found, when large amounts of melting fresh water that was cold upset the saline balance in the water off the Laurentian in this timeframe (the Henry coefficient). For all that, the peak upward right after this, about 7,000 years ago, is HIGHER than arrow 3 on the left (today) with *weaker* downward amplitude. This is consistent with archeologist William Ritchie's "warmer and drier New York State" 7,000 years ago, and most likely, much more floating CO2 and the attendant water vapor.

Figure 2. Close up of Figure 1, showing the 8.2 Kyr event in relation to the Little Ice Age (LIA) and Younger Dryas (11.5 Kyr B.P.).

wind speeds, and a drop in atmospheric Methane levels—Methane being a Greenhouse Gas.[82] A slowing down of the thermohaline circulation as a result of a freshwater flow into the seas has been proposed as the cause of the 8.2 Kyr event. The thermohaline circulation slowdown resulted in a decrease of the northward heat transport in the North Atlantic Ocean leading to pronounced cooling. So freshwater introduced into warmer or salt-dense seawater was such that it affected the flow of ocean currents as perhaps cool winds blowing off of seas. The glaciated Laurentide Lakes in front of the Laurentide ice sheet were most likely the source of the freshwater surge into the salt sea.[83] It could be that melting ice caused a negative climate feedback in ocean circulation, resulting in Northern Hemispheric cooling for those 300 years. What caused the ocean oscillations at this time is not considered here.

A thousand years after the 8.2 Kyr event, or so Figure 2 intimates (and the pseudo-decadal graphs later on in this chapter shows) the Holocene Maximum or "Climate Optimum" occurred. In upper state New York, the archaeologist William Ritchie [84] reported C14 dating had marked the area as "warmer" than today, and perhaps even "somewhat drier," and this is consistent with the approximately two degree Celsius Northern Hemispheric warming that was then ongoing. It strongly suggests that the microclimate of New York State (and of Southern Ontario, Canada) in this part of the Holocene timeframe received a westward-to-eastward invasion of species during the upper peak shown in the Bond graph (arrow 4 in Figure 2). Ritchie hints that New York State's climate and biome was more or less prairie-like:

Two species may record changes associated with the Climatic Optimum [Holocene Maximum in upper New York State] . . . The weather was presumed to be warmer and drier than at present, enabling prairie forms to extend their ranges eastward. This is believed to have been responsible for the prairie mole (Scalopus aquaticus) in an archaeofauna from Pennsylvania, 75 to 100 miles east of its present range.[85]

Based on limited data, the Eastern Fox Squirrel (*Sciurus niger*) and the Eastern Box Turtle (*Terrapene carolina*) also inhabited northern and north eastern New York and Southern Ontario

---

[82] Wiersma A.P. & Renssen H., "Model-data comparison for the 8.2 ka BP event: confirmation of a forcing mechanism by catastrophic drainage of Laurentide Lakes," *Quaternary Science Reviews* (2006) 25, 62-88

[83] Clarke, G., Leverington, D., Teller, J. & Dyke, A., "Superlakes, megafloods, and abrupt climate change," *Science*, (2003) 301, 922-923

[84] Ritchie, William A., *The Archaeology of New York State* (Purple Mountain Press, 1994; reprinted from Bantam/Doubleday, 1965) p. 42. The C14 evidence is derived from pollen samples found at particular undisturbed stratigraphic levels around campsites of the "Lamoka" peoples in New York State. The bones are so deteriorated they could not be racially indentified but a "European" presence not unlike that found at L'Anse Aux Meadows could have been here as early as c. 4,500 B.C.

[85] Quoted by Ritchie (p. 57) from Guilday, J.E., "Prehistoric record of *Scalopus* from western Pennsylvania," *J, of Mammology*, Vol. 42, # 1, pp 117-18, Lawrence, Kansas (1961)

at this time (they do not so any longer except as escapes or sporadic sightings).[86] Indeed, the very movement of the tree line, north, upwards of hundreds of miles near this period has been recorded using isotope proxy and palynological (fossil pollen) data.[87]

Arid areas like prairies (that are also called steppes) are attractive to nomadic or semi-nomadic human hunters due to its favoring by large herding herbivores, as discussed concerning the south-central Siberian steppe hunters of c. 850 before Christ (B.C.) in the section on the "850 B.C. event" below.

As in Siberia in 850 B.C., so in northern New York State in c. 4,500 B.C. Palynological sampling around undisturbed Lamoka people burials in that warmer part of the Holocene "Climatic Optimum" revealed evidence of things that no longer live there. Even the racially-unidentifiable bones of the people themselves were noted by Ritchie, apparently aggressive hunters, and who may have made their way towards what is now upper State New York in the same manner the Vikings did in a similar period of longer-term natural global warming: the Medieval Maximum [88]—only very much later.

---

[86]   Ibid, Ritchie p. 57

[87]   The Ennadai in the Laurentian Shield in my co-authored book, *The Maunder Minimum and the Variable Sun-earth Connection*, p. 215

[88]   Given the primitive state of meteorological and Earth science reconstructing micro climates in antiquity cannot be indentified and measured accurately. The Vikings made settlements in Nova Scotia then and had been—much like the war-scarred bones of the Brewerton Period people revealed thousands of years previously—attacked by the "natives." This perhaps contributed to their abandonment of L'Anse Aux Meadows, Newfoundland.

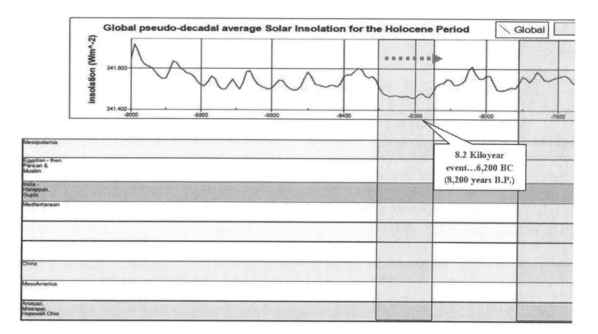

The Hallstatt Cycle in the Holocene: the end of long-term deep minima for 1,000 years?

With such isotopic considerations freshly in mind within deep timeframes (or, "time scales" as scientists like to say) we glance at sunspot reconstruction data into deep time. One such reconstruction is called the Hallstatt Cycle, which takes into account at least 10,000 years of solar activity. For here we introduce a very interesting solar link into deep time climate change.

We know that C14 is valid for at least 60,000 years as a solar activity "counter" as revealed for instance in tree rings (as discussed in Chapter 2). A number of scientists have seen fit to date solar activity backwards into a thick slice of time due partly to this fact. One who could be considered a pioneer in this sense, using the c. 11 Schwabe Cycle sunspot count or C14 or either, was D. Justin Schove, who analyzed such phenomena as long ago as 1955 and for more recent purposes from 200 B.C. as early as 1961. [89] Much more recently however comes the work of Ilya G. Usoskin, Sami K. Solanki,[90] F. Steinhilber, and Jurgen Beer. [91] This work covers

[89]   Schove, D.J., "The sunspot cycle 649 B,C, to A.D. 2000," *Journal of Geophysical Research* 60, 127-156 (1955) and Schove, D.J., "Solar cycles at the spectrum of time since 200 B.C., *Annals of the New York Academy of Sciences*, 95, 107 (1961)

[90]   Usoskin, I.G, *et al.,* "Millenium-scale sunspot number reconstuction; evidence for an unusually active sun since the 1940s," *Phys. Rev. Lett.* 91 (2003)

[91]   Steinhilber, F, Beer, J., and Fröhlich, C., "Total solar irradiance during the Holocene" *Geophysical Research Letters*, Volume 36, 19, (2009)

over 9,300 years—nearly the entire length of the Holocene—and has been further refined since 2009.[92]

Roughly speaking, the Hallstatt Cycle or oscillation describes a long-sloping sine curve that can be broken down into either a positive (less deep minima) or negative (more deep minima) phase. The length of the Hallstatt has been pinned down to either 2,300 years[93] or 2,250 years.[94]

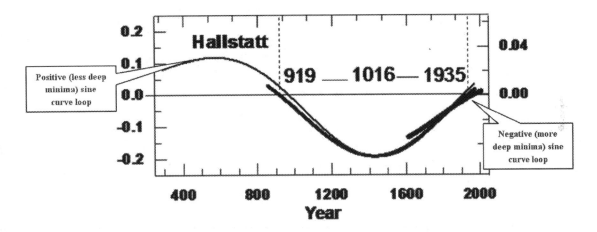

The Hallstatt Cycle from c. 100 A.D. to 2000 A.D.[95]. Shown are three curves from thickest to thinnest: the international; sunspot number, the Schove (1955, 1961) and Usoskin et al.'s (2003) time series, all scaled to fit. The right hand abscissa corresponds with the historical and Schove time series and the left-hand abscissa is that of Usoskin et al. (2003).

In the diagram above, the older data from Schove (1950s, 1960s) the "standard" [96] international sunspot number ("R"), and the quite recent data of Usoskin (2003) describe the positive (from c. 100 A.D. to c. 910 A.D) leg of the Hallstatt and the negative (from c. 910 A.D. to c. 1935 A,D.) Anyway, the upshot of this is that most grand episodes of deep solar inactivity occur after 910 A.D, to include the Oort, Wolf, Spörer, and Maunder Minimums (grand negative episodes) in the negative leg. [97] This could mean that since a positive phase of the Hallstatt began in the year 1935, no major deep solar minima will be witnessed for quite some time; maybe for

92   Steinhilber, F., *et al.* "Interplanetary magnetic fields during the past 9,300 years inferred from cosmogenic radionuclides," *Journal of Geophysical Research* 115 (2010)

93   Cliverd, M.A., *et al,* "Solar activity levels in 2100," *Astronomy & Geophysics*, 44, 5.20 (2003)

94   Ibid Steinhilber, F., *et al*

95   Ibid de Jager, *et al* (in press 2012)

96   More on this number vis á vis recent work by Lockwood is covered in detail in Chapter 7 regarding solar mechanisms.

97   As per Steinhilber

a thousand years or more. What any of this has to do with human activity can be inferred from other diagrams included in this chapter. (More on this cycle and its implications for any potential major global climate change by solar means will be taken up in this book's concluding chapters.)

Powerful regional effects of climate change vs. "global" averaging: three views

To reiterate, solar forcing is very small when compared with what it does to on Earth over time in attempts to get a global average through time. Regional climate effects, connected to ever-increasing knowledge of solar forcing, will probably eventually aid in unravelling the mystery of how Earth's climate functions—or at least how the solar impact to Earth's climate partially functions. Should this ever be known, how humans impact climate might be all that much easier to measure. But to note God's invocation to Job on the difficulties of knowing all the movements of the winds, we can only stand back and wonder.

Regional climate effects are probably in the order of from three to 10 times stronger and then often go in directions completely different from the mean global trend. (This in itself poses many questions.) Averaging them all out yields an overall weak global effect. The weakness of global temperature averaging, as academically useful as it is, is as scientifically suspect as proxy isotopic evidence. Like God to Job in effect, it is a complex "machine" that we do not understand. Still, we take it apart as curious scientists and look at it piece by piece.

Three regional examples of trends from strong local/regional climate-change effects in both cool and warm contexts gives us a striking view of the actual motor of climate change howsoever measured or observed. First shown is cultural change and migrations of peoples in or around 850 B.C., which coincided with an abrupt lowering of solar activity and a consequent degrading of weather which forced cultural change. (Note: this sentence does not imply proof of cause and effect.) The second is the poorly understood Little Ice Age (LIA) which always asks for interpretation and understanding. A third, the effects of locally warmer weather due exactly to this lifting of the "LIA" on avian migration and resettlement or pioneering behavior, is next.

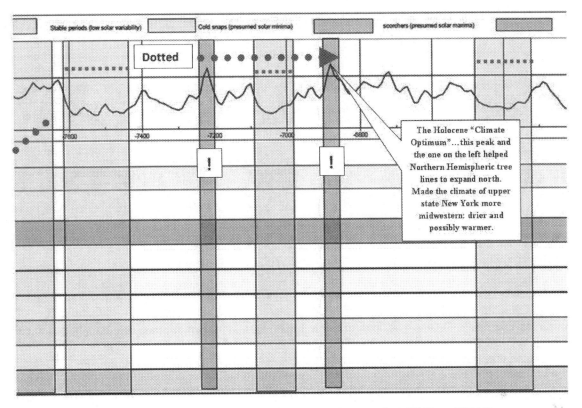

Sliding frame data showing the Holocene Maximum (after Bill Howell:2012)

One: "the 850 BC event": lower solar activity with higher precipitation initiates a human crisis in the Netherlands and aridity in the tropics: changing populations, changing cultural behavior? [98]

There was a degrading of weather that was noted in proxy data drawn from peat bogs about 850 years before Christ in the Netherlands. Climate-related changes in precipitation and temperature are reflected in the changing species composition of peat-forming vegetation as evidence for this fact.[99]

Plant remains are infinitely identifiable. By using ecological information about peat-forming species, changes in species' composition of sequences of peat samples can be interpreted as evidence for changing local hydrological conditions linked to climate change. At the start of the abrupt climate shift (see Figure 3, middle)—and coincident with an abrupt decline of

---

[98]   The "850 BC event" is detailed in "Holocene climate change and the evidence for solar and other forcings" by Jürg Beer and Bas van Geel (2008)

[99]   Ibid Beer, J., and van Geel, B., (2008)

solar activity—the atmospheric circulation changed, leading to cooler and wetter climate conditions.

In lowland regions in the Netherlands for example, the climate shift caused a sudden, considerable rise of the groundwater table so that arable herding land turned into wetland, where peat growth then started, as peat favors such wet conditions. Farming communities living in lowland wet areas were forced to migrate because they could no longer produce sufficient plant and animal food. [100]

The rise of the water table forced farmers to migrate to well-drained areas in the northern Netherlands where salt marshes offered them new fertile land. (Phase C and B, top and middle in Figure 3). Notably, the sequence begins roughly in 950 B.C.—a warm period near the time Vikings explored Greenland. The rise of inland water tables is attributed to increased precipitation. (becoming part of the sediment and so, collected by researchers as core samples). But by 750 B.C., warming returned (Phase C, top of Figure 3).

The Netherlands was not alone in regional cooling near or at this time. Evidence from proxy data in this timeframe also suggests climate cooling events in France, Switzerland, Central Russia, and in the Andes in South America. All these latter-mentioned were seen due to palynological evidence revealing vegetation shifts consistent with global cooling. There is also evidence for dryness in Central Africa and Western India. It was shown [101] that over a period of several millennia the presence of lakeside villages in Southeastern France and adjacent Switzerland was strongly linked with lake levels and solar activity. Lakeside villages were present during periods of high levels of solar activity, as evidenced by reduced atmospheric C14.[102] As C14 production is regulated by solar activity, periods of increased mire surface wetness and increased lake levels (peaks of C14) have been interpreted as evidence for solar forcing of climate change (the effects of sudden declines in solar activity).[103] No lakeside villages were maintained after 850 B.C. here.

---

[100] van Geel, B., Buurman, J. & Waterbolk, H.T. "Archaeological and palaeoecological indications of an abrupt climate change in The Netherlands, and evidence for climatological teleconnections around 2650 BP." *Journal of Quaternary Science* (1996) 11(6), 451-460

[101] Magny M., Holocene climate variability as reflected by mid-European lake level fluctuations and its probable impact on prehistoric human settlements." *Quaternary International* (2004) 113, 65-79

[102] And more of it heading for trees and other biota

[103] By wiggle-matching C14 measurements, high precision calendar age chronologies for peat sequences can be generated (Blaauw *et al.* 2003), which show that mire surface wetness increased together with rapid increases of atmospheric production of C14 during the early Holocene, the Sub-boreal—Sub-atlantic transition: a sharp increase of C14 production and evidence for wetter conditions and the LIA (also Wolf, Spörer, Maunder, and Dalton minima of solar activity).

A link between the climate shift around 850 B.C. and the evidence for a subsequent increase in human population density has been made in north western Europe.[104] A climate crisis in the first instance caused environmental and social upheaval. A collapse of societies resulted in a weakening of the position of dominating groups, which brought about a change in the social structure of farming communities. This facilitated the introduction of a new technological complex, which again created further social change combined with a leap forward in production, food consumption, and population density. In this case there was apparently no catastrophic decline in human existence, but a major disruptive shift due to climate drivers for the cooler.

In south-central Siberia near this time archaeological evidence suggests an acceleration of cultural development and a sudden increase in density and geographic distribution of the nomadic Scythian population after 850 B.C. It was then hypothesized that a relationship with an abrupt climatic shift towards increased humidity (equator ward relocation of mid-latitude storm tracks) such as that driven by the Hadley Cell. The hypothesis is supported by pollen-analytic (fossil pollen, or palynological) evidence. Areas that initially may have been hostile semi-deserts changed into attractive steppe (alternatively called prairie) landscapes with a high biomass production and carrying capacity. Newly available steppe areas could be utilized by herbivores, making them attractive for nomadic tribes. Compare this with the aforementioned case of the Lamoka people in what is now upper State New York in about 6,000 B.C. who apparently were fierce hunters and some of whose stone tools are known as the Brewerton type. The Central Asian horse-riding Scythian culture expanded and an increased population density was a stimulus for westward migration towards south eastern Europe.

There is strong evidence for climate change in the Central African rain forest belt around 850 B.C. as well. [105] Palynological studies point to a drastic change in the vegetation cover (from predominantly rain forest to a more open savannah landscape) as a consequence of wide scale dryness not necessarily connected to destructive drought but to a gradual warming. A population of farmers migrated from the south into the area. The contrast between this change to dryness in central west Africa and the contemporary increase of precipitation in the temperate zones fits well with the hypothesis that, after a decline in solar activity, there was a decrease in the latitudinal extent of the Hadley Cell circulation and consequently, mid-latitudinal monsoons decreased in intensity. Meanwhile, the mid-latitude storm tracks in the temperate zones were enhanced and moved in the direction of the equator. [106] Cold can bring unexpected destructive floods or droughts as much as can heat.

---

[104] van Geel B., Bokovenko, N.A., Burova, N.D., et al. "Climate change and the expansion of the Scythian culture after 850 BC: a hypothesis." *Journal of Archaeological Science* 31(12), (2004) 1735-1742

[105] van Geel B., van der Plicht, J., Kilian, M.R., et al. "The sharp rise of Δ14C ca. 800 cal BC: possible causes, related climatic teleconnections and the impact on human environments." *Radiocarbon,* 40(No. 1), (1998) 535-550.

[106] Van Geel, B. & Renssen, H., "Abrupt climate change around 2650 BP in North-West Europe: evidence for climatic teleconnections and a tentative explanation." In: *Water, Environment and Society in Times of*

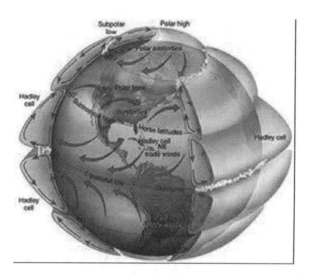

The Hadley Cell. Declining solar activity interrupts this cell's circulation causing decreasing monsoons. Yet storm tracks in the temperate zones increase at such times, forcing degraded weather towards the equator. As droughts increase in some "local" areas where monsoon usually dominates in low solar periods, does flooding occur "locally" (and unexpectedly) elsewhere? (Carleton University)

We note that these proxies show a hemispheric trend. A drought caused by a weak monsoon intensity in north western India after 850 B.C. also supports this hypothesis. [107] Moving toward the Americas, massive glacier advance in the south-central Andes of Chile, probably resulting from an equator ward relocation of mid-latitude storm tracks forms part of a wealth of evidence for "worldwide" climate change around 850 B.C. for the cooler.[108] Evidence from paleodata indicates that the climate shift around 850 B.C. occurred suddenly, probably within a mere decade, and (at least [109]) the C14 record points to a sudden, Maunder Minimum-like decrease of solar activity as the cause of this event (see the pseudo-decadal averaging chart for 1600 BC-400 B.C. below and the location of the dashed arrow).

The theory with the close-up of this 850 B.C. event, stepping back and taking a long view culturally, is noting solar weak periods and the rises and falls of civilisations. The theory loosely considers not just temperature, but perhaps more like precipitation, and not only warfare indices. Plagues and other climate disruptors are dwelled on. For example, heavy cloudiness as

Climatic Change. Eds. A.S. Issar & N. Brown. Kluwer Academic Publishers, Dordrecht (1998) pp. 21-41

[107] Van Geel B., Shinde, V. & Yasuda, Y., "Solar forcing of climate change and a monsoon-related cultural shift in western India around 800 cal. yrs BC." In: *Monsoon and Civilization,* Eds Y. Yasuda & V. Shinde, Roli Books, New Delhi. (2004) pp. 275-279

[108] Van Geel B., Heusser C.J., Renssen H. & Schuurmans C.J.E., "Climatic change in Chile at around 2700 BP and global evidence for solar forcing: a hypothesis." *The Holocene,* 10(5), (2000) 659-664

[109] Be10 and others are also "counters."

a negative feedback of say, albedo from widespread and persistent lower tropospheric cloud build up in key seasons that can reflect up to 90% of the Sun's short-wavelength (visible light). Attendant precipitation shifts from very high to low and resultant droughts are others. Floods and the crop failures connected with these as well as with crop diseases, insect "plagues" (such as locust clouds), agrarian society economic failure (as just seen in the well-documented "850 B.C. event") and pandemics also factor in. These connect to each other in some cases, possibly as wars are prompted due in part to crop failures and famine-forced diseases that partly are the result of a weakened Sun wreaking havoc on crops and, by proxy, markets both wide and local. There is a feeling that the zones of huge agricultural productivity wander with the longer term solar cycles and even short-to-mid-long term pseudo cycles. As solar insolation decreases, big regional effects differ in nature. To quote Bill Howell, "There forms an intriguing dichotomy between 'desertification' versus 'junglification' in some respects to witness the rather sudden shifts from dryness to wetness, and the attendant changes in forces on humans."

The end of the 850 B.C. event, however, seems to have been gradual (a time-transgressive passing of thresholds) so that, given present knowledge, it is not yet possible to pinpoint an end of the event. The changing climatic conditions at 850 B.C. may have been similar to climatic cooling shifts during the LIA.[110]

Two: the "Little Ice Age" (LIA)

Sandwiched in between the Bond graphs (see Figure 2) in tight amplitudes, barely visible, from the mid-1500s to the end of the 1800s is the phenomenon labeled such that it confuses most who stumble upon the subject. That would be the anomalous Little Ice Age or LIA as it is technically referred to. It was caused either solely by the Sun / albedo etc., solely by geophysical events (volcanoes etc.) or was purely an ocean (hydrological) phenomenon. It may have been caused a little by all of these and it defies a neat, cyclical cubbyhole. In any case, as shown in Figure 2 it cannot possibly be confused with a deep ice age. The naming of the LIA was quite accidental and innocent. At the time it was labeled it made great sense however.

The term "Little Ice Age" was coined by a journalist, probably in the 1930s according to US Geological Survey scientist F.E Matthes (in 1940) as he described then-occurring [111] glacial re-expansion in a post-Pleistocene context on page 398 of an AGU report:

They (the glaciers) have re-expanded since then to the limits from which they are even now receding, and as their re-expansion has been of considerable magnitude, to judge from certain specific cases, there appears to be a warrant for the assertion that the present age is witnessing

---

[110]   Mauquoy, D., van Geel B., Blaauw, M. & van der Plicht, J., "Evidence from northwest European bogs shows 'Little Ice Age' climatic changes driven by variations in solar activity," *The Holocene,* 12(1), (2002) 1-6

[111]   That is, in the 1930s

a mild recrudescence of glacial conditions—that it is, as a clever journalist has suggested, a separate "little ice age."[112]

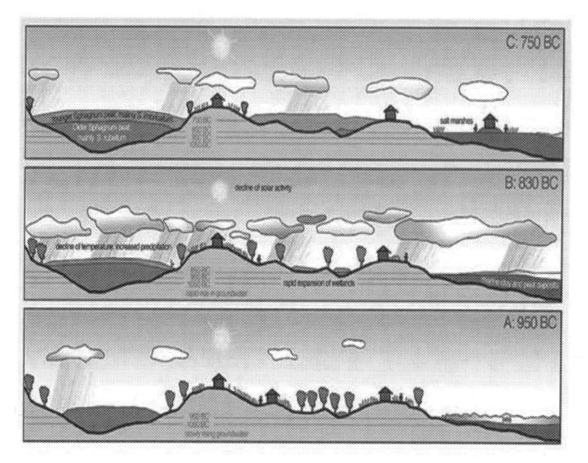

Figure 3. Three views of climate change around 850 BC as revealed by increases in peat bog growth in the Netherlands. Phase A (base), warm and dry. Phase B (middle; c. 850-730 BC) cold and wet. Phase C (top) a return to warmth near 750 B.C. (After Beer & van Geel, 2008)

What nags at the understanding of the LIA, other than the confusing label, is its locus in the range of two well-known solar minima: the Spörer (possibly a grand solar episode) and the Maunder. (definitely a grand solar episode). If separate from any solar activity, the LIA may have worsened the already-rugged climate conditions in the Northern Hemisphere at the time. We can see that from the familiar Eddy graph of C14 per mille and the curve in the graph below. In

---

[112] Matthes, F.E., "Committee on Glaciers, 1939-40." *Transactions, American Geophysical Union, 1940* pp. 396-406. Parenthetical information supplied by this author.

any case, the LIA's end coincides with solar insolation increases overall since Solar Cycle 11 [113] or so, and could be one of the contributory rebounding effects to a warmer 20th Century—very much so after the year 1924, at least according to this book's thesis.

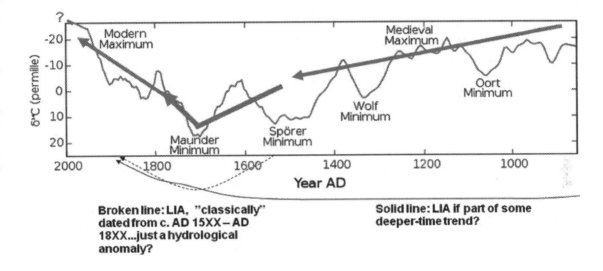

Broken line: LIA, "classically" dated from c. AD 15XX – AD 18XX...just a hydrological anomaly?

Solid line: LIA if part of some deeper-time trend?

Exacerbating the effects of prolonged solar minima was the coincidence of two closely-occurring grand minima episodes one after the other. This is covered elsewhere in detail.[114] Weather in Europe had already been generationally different from what Spörer and Maunder Minimum-living old timers recalled, which fell into what some think was the coldest year in eight thousands, culminating at the end of the year A.D. 1683, once called the hardest freeze (of the Thames River in England) in postglacial times [115] (that is, prior to c. 10,000 years BP). But from the Bond graph in Figure 2, the hardest looks closer to c. 8,000 years before the year A.D.1683.

Three: local shorter term warming since 1880 (post-LIA) in the altered permanent residency of migratory birds northward

We now do a hyper-convergent thing and make a tie-in to the science of ornithology in this largely anthropomorphic examination of climate steered events. Rather than think this a digression, it is a convergence from other branches of science used to lend weight to any branches of science attempting to understand natural local shorter term Earth warming or cooling. The following explanation purports to show that species of birds—as warm-blooded

---

113   A full list of solar cycles from c. cycle 10 is supplied in the concluding chapter of this book.

114   Ibid, Soon, W., and Yaskell, S.H., *The Maunder Minimum and the Variable Sunarth Connection* (WSP:2004)

115   By postglacial it is probably meant after the so-called *deep* ice age's end. For it is not clear if the ice age is quite over as of yet. (Ibid, Soon-Yaskell, 2004)

as mammals if not more so—have moved permanently northward since the end of the LIA, and which is a regional—if not hemispheric—phenomenon.

A never ending anecdotal series of litanies on how much colder it was in the "old days" in the US Northeast (roughly 100 years ago) and somewhat earlier wends its ways through popular literature. New England itself has been proposed to have odder weather due in part to its peculiar lithospheric position contra the atmosphere.[116] But some of these litanies are more accurate than supposed. This period is also conjured as a foil to dramatically contrast with current or recent apparent hemispheric warming by humans alone, depending on how you date massive CO_2 release into the atmosphere by people. A scientific book on Massachusetts birdlife [117] as it applied to state agriculture in 1905 supports the notion of permanent avian northward settlement. It was carefully assembled by numbers of good observers reporting to a professional ornithologist (E.H. Forbush) and relates later times for spring arrivals of many species, and earlier migrations of them south and west in the fall back then compared to notes and data in modern field guides. Some species common in Massachusetts today were rare there in 1905 (like the Tufted Titmouse [*Parus bicolor*] and Cardinal [*Cardinal cardinalis*]) or never occurring (like the soft-footed Mourning Dove [*Zenadoura macroura*] which became a year-round resident in Massachusetts in the 1940s) and the House (or Mexican) Finch (*Carpodacus mexicanus*). The Mexican Finch arrived in the US from the south around 1940 into Texas [118] and was spread from Long Island around 1950.[119] It has been a common summer and fall bird in Massachusetts for some years now. Many birds listed as seasonal residents in the 1905 book are now year round residents there today. It must have been cooler locally before c. 1900 in Massachusetts compared to after that time (c. 1950). Resource abundance must have brought them northward on wings, the possible help of human relocation notwithstanding, and warmer years making them year-round residents in the higher north.

Another issue in this physical transformation could be the relative strength of the magnetic field. This northward permanent migration of passerines must have been ongoing since the 1860s if not earlier. Perhaps strengths and weaknesses in Earth's magnetic field signal the passerines to fly farther afield—north or south—depending on the signal strength they receive

[116] Herman, J.R., and Goldberg, R.A., *Sun, Weather and Climate* (NASA, SP-426, 1978)

[117] Forbush, E.H., *Useful Birds and Their Protection* Mass. State Board of Agriculture Publication (Wright & Potter: 1905). Forbush was the official state ornithologist for Massachusetts then.

[118] Peterson, R.T., *Birds of Eastern and Central North America* (Houghton Mifflin: 2002)

[119] Sibley, D.A., *The Sibley Field Guide to Birds of Eastern North America* (Knopf: 2003)

in their olfactory (breathing) glands.[120] [121] In any case, their increased migrations northward for longer periods started happening before the intense and widespread use of fossil fuels among an Earth population of less than two billion humans (pre-1940). Again, this point is widely debated.

Pseudo-decadal averaging solar insolation viewed in context with human population
An overview of civilization rises and falls matched to variations in solar insolation from 1,600 B.P. to present

How does one characterize "periods, phases or states" of a chaotic system (such as solar variability) on another chaotic system—the "system" of the Earth's climate? This question is one of intense interest in the field of solar science, and increasingly so in geophysics, in aeronautics, and in space research.[122] It even has ramifications for the study of human demographics and conflict issues, this latter referred to as peace studies. The survey here is in no way intended to imply environmental determinism: nor is the opposite of planned and deliberate control of human societies for "human safety" in the face of any natural hazards implied. Shown here is the middle view.

The following tables portray averaged peaks (341.600 Wm²) in solar insolation and dips (341.400 Wm²) from 1,600 years ago to the present time.[123] The tables bear some study and serious

---

[120]   It is also interesting to note that passerines are now known to have an organ that can detect the Earth's magnetic field for navigation. Bobolinks (or, Ricebirds) *Dolichonyx ozivorous*, common in the US east, have Iron Oxide around their olfactory glands and in bristles extending from naval cavities. "Closely proximal nerves fire in response to changes in magnetism to provide the exquisite sensitivity necessary for navigation by the Earth's magnetic field." ("Close Up Science," *Valley News*, Vermont, July 25, 2011)

[121]   Additionally, can passerine motion teach us something about wind dynamics we do not know about? Italian theoretical physicist Giorgio Parisi tried to explain a wide-scale flocking behavior called a "murmuration" performed by a common trans-Northern Hemispheric bird, the Starling (*Sturnus vulgaris*). It mimics wind motion and velocities. According to writer Brandon Keim on Parisi, "the math equations that best describe starling movement are borrowed from the literature of criticality, of crystal formation and avalanches—systems poised on the brink, capable of near-instantaneous transformation. They call it scale-free correlation, and it means that no matter how big the flock, If any one bird turned and changed speed, so would all the others." *(The Atlantic:2011)*

[122]   Valdes, B.C., *Time Dependent Neural Network Models For Detecting Changes Of State In Earth and Planetary Processes*

[123]   Based on Laskar *et al*, and Solanki, Tapiping. Laskar solar insolation results assume 1368 Wm^-2. (1) Likely error that the insolation curve is shifted 85-years to left due to misinterpretation of Solanki's data table. (2) The sunspot variances as shown through insolation have been exaggerated to show what the Hale 11-year half-cycle peaks and troughs would look like (no galactic ray-cloud impact). This would skew energy perspectives, but may give an idea of shorter-term shocks as experienced by civilizations. Perhaps the temperatures are mostly influenced by GLOBAL irradiance iariations (and leveraging such

consideration as regards society, sickness, war, etc., and the Sun's relative strength. (Though droughts, plague and war etc. occur along with the mapped out solar insolation we do not imply that the Sun is the sole cause of the listed droughts, plagues, and wars.) The two-way arrows show the rises of various civilizations to their respective terminus points and then peter out into newer periods, or lapse into take over by other civilizations. From the top down is shown the Mesopotamian, Egyptian, Indian, Mediterranean, European, Chinese, and Mesoamerican/Anasazi-Mississippian-Woodland societies on many of the world's very-peopled landmasses.

The dotted over areas are periods of presumed stable solar maxima, the dashed are the presumed solar minima, regardless of their strengths or whether or not they were grand episodes. Exclamation marks indicate "scorching" warm periods.

Several things stand out in a short inspection of the graphs. One is a confluence of arrows (loosely interpreted as population disruption or change or both) in areas between 800 B.C. (and the 800 B.C. Minimum) heading into the weaker maxima periods found around 600 to nearly 400 BC. From the work of van Geel and colleagues gone over earlier in this chapter, it is amply clear that during such minima there was considerable civilization disturbance (as he and colleagues outline well) in the Netherlands and elsewhere.

In the 800 B.C. timeframe the Phoenician and Etruscan civilizations transform into the earliest parts of the Roman Empire, this famous empire in full swing by 509 BC, or three hundred years subsequent to 800 B.C..[124] The rise in European population falls off after the Roman Empire's fall by around 300 after Christ's death (or A.D). In the relatively warm period stretching from c. 550 A.D.-850 A.D, Charlemagne encouraged the rise of feudalism and it can be said that this labyrinthine system of vassal and sub-vassaldom was popular enough to take root and help propel (from about A.D. 700) at least mid-Europe's population from c. 23,000,000 to upwards of 73,000,000 humans by the year A.D. 1250 (through the Oort Minimum) and into the Medieval Maximum. This rise is the first such major reversal in human population growth since late Roman Empire [125] times in this Northern Hemispheric sector (at least). For human populations had been sinking steadily in that sector up till then since the Roman Empire's main collapse.

---

as galactic rays-clouds, water vapor-clouds, ice albedo, ocean absorbance, etc.), whereas crops are also greatly affected by seasonal variations.

[124] Chronologies of ancient history prior to roughly the time of Alexander the Great in terms of time series analysis of war are here, with warning added, highly suspect. Conventional dating in a multiple conflicting hypothesis approach for conflicts going back into very deep time are held onto strictly for pedegogical reasons. (Private communication, Bill Howell, April 2012)

[125] Harrison, J.B., and Sullivan, R.E., *A Short History of Western Civilization, Vol. 1: to 1776* 4th Ed. (Knopf: 1971) pp 208-209 for these figures and analyses. These in turn were drawn perhaps from Georges Duby, *Rural Economy and Country Life in the Medieval West* (Harvest)

The geographical center of this upward surge seems also to have been in what is today western Germany, north, and west to the Netherlands, the Rhine River region, then south to northern Italy. History textbooks relate that reasons for this period of massive population growth of c. 10 million persons per century for five centuries in a row remain obscure. That the rise coincided with massive agricultural production in west-central Europe, southward along the Rhine (prompting that very political yet mild growth, feudal and manorial life) is of course not so surprising. But, what should *also* not be so very surprising is that agricultural yield and population growth at this critical juncture both coincide with the Medieval Maximum which was very kind, climatically-speaking, to Western Europe. Curiously, North America, in the 550 A.D.-600 A.D. period, and again in the c.1050-1250 A.D. timeframe is burned up by drought, at least in the North American west.[126]

In the 400 B.C. Minimum to nearly 200 B.C, the Persian Seleucid civilization is taken over by the aggressive and world-changing Roman civilization. Eastward, the Maurya Dynasty in India dies out into the Gupta, and, following a weirdly even pattern from then on, Chinese dynasties routinely die out, and subsequently re-form, in each solar minimum. This odd pattern begins with the Chinese post-Zhou interregnum (403-221 B.C.), goes into the Chinese so-called "Dark Ages" of 220 BC-581 A.D. (the 200 A.D. Minimum) all the way to the start of the Sui Dynasty (581 A.D.-618 A.D.) to the Tang Dynasty's demise nearing the Oort Minimum and the beginning of the Mongol Empire, including the Yuan Dynasty, to the Ming Dynasty's beginning in the Spörer Minimum. The Ming dies out in the even worse (LIA-aggravated?) Maunder Minimum, whence it becomes the Qing (Ch'ing) Dynasty. The human suffering in the Chinese / mainland Asian context and framework in this particular period is well outlined elsewhere and bears reconsideration in light of this pseudo-decadal solar insolation data by Howell. [127]

---

[126]   The pioneering dendrochronological work of A.E. Douglass delineated this dessication from his western US tree ring counts from trees living in the US in the 1200s.

[127]   Ibid, Soon-Yaskell (2004)

1600 BC–400 BC (slide window)

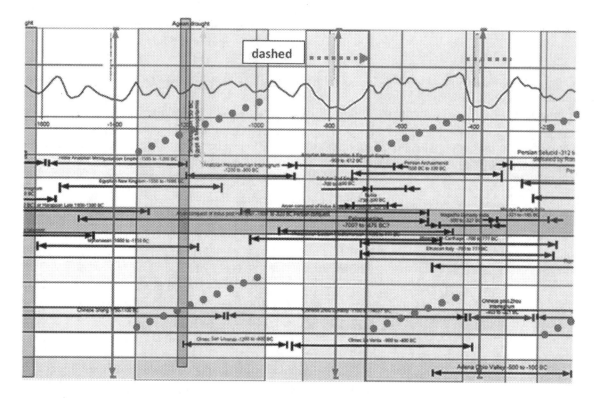

Mesoamerican experiences show a similar pattern to the Chinese, in that the Hopewell people of the Mesoamerican Ohio Valley thin out in the A.D. 200 Minimum, the Anasazi (US southwest/ northern Mexican) civilizations rise at the very end of the Oort Minimum and fade out into the Wolf Minimum, and what Spain and Portugal wrought in this part of the world (to include, most prominently, what occurred to the Aztecs) is well-known and documented, the Spanish tide itself losing steam in or around the Dalton Minimum.

400 BC—c. 1300 AD (slide window)

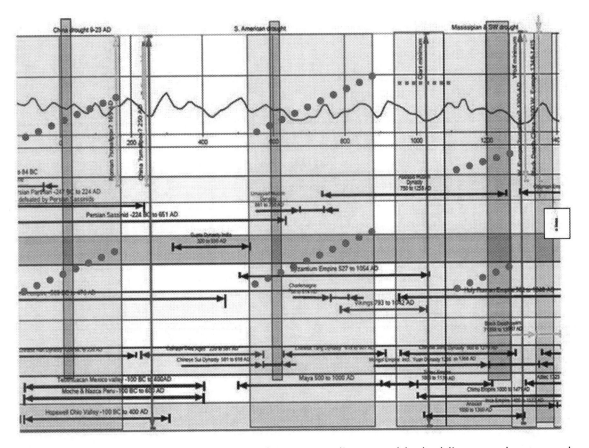

Taken on rough average, the amounts of arrows ending at a blocked line are about equal, whether or not we are looking at climate optimum periods (dotted lines) or the climatically cooler, drier, windier ones across the 1,600 B.C.—present timeframe. But a skew becomes apparent if we look at those civilizations from the northern Mediterranean upward to northern Europe, including middle China and North America, versus the more southerly to Southern Hemispheric-occurring civilizations.

c. 1300 AD—c. 1950 AD (slide window)

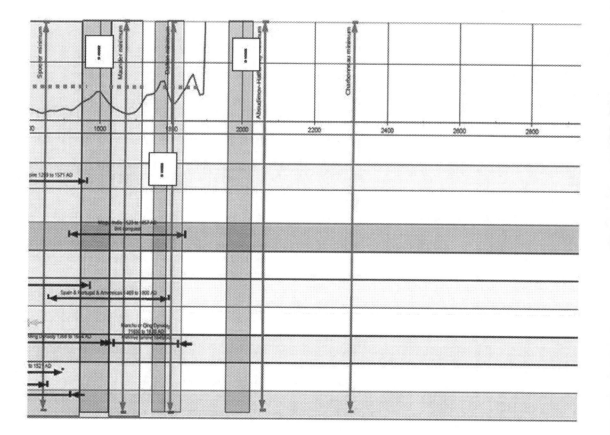

Solar minima and recurrence of civilization collapse, northern hemisphere relative to the southern hemisphere	
1400 BC Minimum	1 of 3 in the Northern Hemisphere
800 BC Minimum	1 of 3 in the Northern Hemisphere
400 BC Minimum	3 of 6 in Northern Hemisphere
200 AD Minimum	5 of 6 in the Northern Hemisphere
Oort Minimum	All 4 in the Northern Hemisphere
Wolf-Spörer-Maunder Minima	5 of 7 in the Northern Hemisphere
	30 total: majority in N. Hemisphere (19 out of 30)

When talking of drought, on the other hand, another pattern emerges:

Recurrence of civilization collapse in drought periods, not minima-dependent, northern hemisphere relative to the southern hemisphere	
Aegean Drought	2 of 6 in the Northern Hemisphere
"600 BC" Drought	1 in 6 in the Northern Hemisphere
China Drought	1 in 3 in the Northern Hemisphere
S.American Drought	3 of 6 in the Northern Hemisphere
Mississippian and S.W. U.S Drought (1200s)	2 of 3 in the Northern Hemisphere
	24 total: majority in the Southern Hemisphere (15 out of 24)

Conflict and climate: statistical studies in a very recent context (1950 AD-2004 AD)

Societal "rise and fall"—a blanket term for a myriad amount of human demographic / geopolitical fluctuations punctuated often by conflict, characterised even more by regime shifts, changes in resource-achieving and maintaining methods and gambits and so on—concerns social and political science study as well as peace research. These "hard" sciences (harder than astrophysics in their grappling with unfathomable, quickly-shifting complexities which answer to few physical "laws") make their ways to the apron of the stage of climate change via the back door of the environmental sciences. These in turn rise up to meet geophysics and solar astrophysics at the confluence point of a rabidly-discussed topic: the strength of the El Nino Southern Oscillation (ENSO) and its effect on climate. ENSO is itself regionally created, but is hemispherically and globally far reaching.

A recent statistical study has shown a correlation between increased conflict and climate change.[128] (The methods used are summarized here. [129]) The introduction to the research letter [130] in the same issue of *Nature* describes the study in the simplest terms:

Hsiang and colleagues' study [the researchers involved] proceeded in two steps. In the first step, the authors used historical climate data to divide the countries of the world into two groups: 93 'teleconnected' countries which have strong ENSO-related climate effects, including Australia, Ghana, Laos, Sudan and Trinidad; and 82 non-teleconnected ones that don't experience these effects, such as Afghanistan, Greece, Latvia, Sweden and Tunisia . . . In the second step, they used statistical models to see whether the rate of outbreak of civil conflict each year from 1950 to 2004 correlated with an annual index of ENSO for the two groups.

The result was as follows, according to the author who analyzed the research letter: [131]

The analysis identified a statistically significant relationship between the rate of outbreak of conflicts and ENSO among the countries in the teleconnected group, but not among the others. In the teleconnected group, the rate of conflict increased from an estimated 3% in La Niña [that is, cooler water surface [132]] years to an estimated 6% in El Niño years [that is, warmer water surface years].

The paper's authors identified what they term Annual Conflict Risk (or ACR). They concluded that: using data from 1950 to 2004, we show that the probability of new civil conflicts arising throughout the tropics doubles during El Nino ["warmer event"] years relative to La Nina

---

[128]   Hsiang, S.M., Meng, K.C., Cane, M.A., "Civil conflicts are associated with the global climate," Research Letter, *Nature*, Vol. 476, 25 August, 2011 pp 438-441

[129]   "Pixels with surface temperatures significantly and positively correlated with NINO3 for at least 3 months out of the year are coded 'teleconnected'; remaining pixels are coded 'weakly affected'. Countries are coded teleconnected (weakly affected) if more than 50% of the population in 2000 inhabited teleconnected (weakly affected) pixels. Group-level time-series regressions (Table 1, models 1-4) use a continuous variable for ACR; we drop 1989 because it is a 3s outlier, presumably because of the end of the Cold War. Group-level standard errors are robust to unknown forms of heteroscedasticity. Country level longitudinal regressions (models 5 and 6) are linear probability models for conflict onset with standard errors that are robust to unknown forms of spatial correlation over distances no more than 5000 km, serial correlation over periods no more than 5 years and heteroscedasticity 21. We estimate the number of conflicts associated with ENSO by assuming all conflicts in the weakly affected group were unaffected and a baseline ACR of 3% for the teleconnected group would have remained unchanged in the absence of ENSO variations. We then project the observed sequence of NINO3 realizations onto our linear conflict model dACR/dNINO350.0081) and find 48.2 conflicts (21%) were associated with ENSO." (Ibid Hsiang *et al*, p 440)

[130]   Solow, A.R., "Climate for Conflict," *Nature*, Vol 476, 25 August, 2011 pp 406-7

[131]   Ibid, Solow

[132]   Parenthetical inserts by this author

["cooler event"] years. This result, which indicates that ENSO may have had a role in 21% of all civil conflicts since 1950, is the first demonstration that the stability of modern societies relates strongly to the global climate.

What social and political scientists—in or out of peace studies—see in this data as cause and effect in human societies is up to them.

A tie between cyclonic activity—which ENSO very definitely is—and solar rotation, is century old information. E.W. Maunder noted this in a paper outlining data collected by British watch stations in their former empire and connecting these to solar rotation.[133] In addition, he posited a connection between solar "stream lines," sunspots, and cyclonic activity.

That the El Nino Southern Oscillation (ENSO)—a primary component of global climate [134]—is perhaps one of the strongest motors of it and is a powerful factor in Indian Ocean monsoons, among others, is no mystery. "The coordinated El Niño/Southern Oscillation phenomenon (ENSO) is the strongest source of natural variability in the global climate system" [135] goes one claim and it is barely exaggerated. Its spread and affects, worldwide, out of the lower Pacific Ocean, was the main theme of that study tying societal conflict to ENSO intensity. [136]

El Niño is the warmer apparition of this massive cyclonic eruption in the southern Pacific Ocean. La Niña is the cooler variant, affecting surface sea temperature. Taken together, these two comprise "ENSO." This odd warming (El Niño) or cooling (La Niña) of surface water in the eastern equatorial Pacific occurs at irregular intervals between 2 and 7 years in conjunction with ENSO. ENSO is "a massive see-sawing of atmospheric pressure between the southeastern and the western tropical Pacific." [137] El Niño was seen as the appearance of unusually warm water in the Pacific Ocean starting near the beginning of the year and in Spanish might refer to the Christ child (Jesus) since the phenomenon arrives around Christmas (December 25). La Niña means The Little Girl. La Niña and means "a cold event" or period. El Niño on the other hand is known as "a warm event."

---

[133] Such as by the earlier-noted General Edward Sabine. Maunder, E.W., "Notes on the Cyclones of the Indian Ocean," *Monthly Notices of the Royal Astronomical Society (MNRAS)* November, 1909. pp 46-62.

[134] It is also tied to solar activity. See Theodor Landscheidt's published papers, for example, Landscheidt, T. (2000 a): *Solar forcing of El Niño and La Niña.* European Space Agency (ESA) Special Publication 463, 135-140. Also earlier, and published online (and so, more controversial): "Solar Activity Controls El Niño and La Niña," Schroeter Institute for Research in Cycles of Solar Activity Nova Scotia, Canada (1999)

[135] Philander, S. G. H., *El Niño, La Niña, and the Southern Oscillation.* San Diego, Academic Press, 1990, as quoted in Landscheidt (1999)

[136] That is, Hsiang

[137] Landscheidt (1999)

The late Theodor Landscheidt pointed out (1999) that "Small Finger Cycles" (or SFCs) of 7.5 years are associated with Southern Oscillation (SOI) extreme patterns between 1951 and 1998, ENSO being this southern oscillation in toto:

After the Big Finger Cycle [BFS—which are shorter than the 11 year Schwabe Cycles] of 1968 they . . . are closely correlated with negative extrema (El Niños [warm]) and before 1968, with positive extrema (La Niñas [cold]). Initial phases of Special Small Finger Cycles (SFS g) are also indicated, but only tentatively and only after 1968. Before 1968 they do not show the usual reversed pattern, or better, there is no consistent pattern at all. This speaks against the dependability of this factor, at least in the period 1951-1998. This all the more so as the two cases where there is a coincidence with El Niños can as well be explained by 0.382 year distance within subcycles of the sunspot cycle.

Objections to the solar link in climate change

The Hsiang paper shows a conflict rise due to global climate change trends from ENSO. If this is accepted, then by proxy, actions of the Sun also influence ENSO's severity in light of the heightened cyclonic rate-to-solar-rotation/ENSO data gathered so long ago. As the Hsiang study's sponsor rounds the thought out, "bitterly cold winter weather contributed to the failure of France's invasion of Russia in 1812. Hsiang et al now report that global climate can be a cause of war." [138]

That Nature can steer us so takes us back to the original thesis of this chapter, and to this book's aim, in general, overall. The objection to Nature being at least a contributory cause of war in *any* dynamic—for the warmer or the colder—is now being challenged by quantitative analysis from the fields of conflict studies and peace research (with its handmaiden of political science). That this should drive physical scientists harder in their search for cause and effect from natural causes in climate change, is lauded as much as those from purely human-issued sources of climate change. People of course determine and wage wars: not the Sun or climate. What is not ruled out however are climate concerns—whether solar-induced with human assistance or not—as an ancillary driver of armed (and so, expensive and destructive) human conflict.

---

[138]   Ibid, Solow p 406

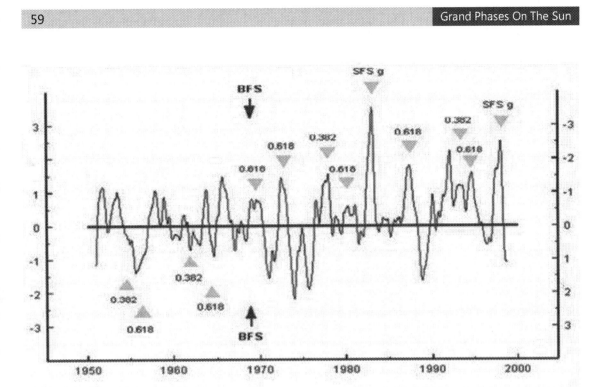

Figure 4. Smaller "finger" cycles of solar activity are common from 1950 until 1968. After that, and a "big finger" imprint, they are more common. Sunspot cycles, then, had apparently been the "bearers" of increased El Nino—"warmer events"—since 1968. (After Landscheidt [1999])

That the Sun (as part of Nature) conspires implicitly in this arrangement of death and destruction (again, generally) is still suspect of course. Landscheidt comments on the suspect case: [139]

Taken together, the lines of evidence presented . . . leave little doubt that the relationship between phases . . . within solar cycles and ENSO events is real. Nonetheless, it is to be expected that sceptics [in this context, those who doubt a *natural* footprint, vs. a purely *human* one [140]] will point at the lack of detailed cause and effect arguments and properly quantified physical mechanisms. Seen in a historical light, such objections are not valid. *The lack of elaborate theory does not impair the heuristic importance of the results. You cannot achieve everything at the same time.* [141] Epistemologically, the stage of gathering data, establishing morphological relations, and setting up working hypotheses necessarily precedes the stage of elaborated theories. How can we solidly connect solar activity with climatic change as long as neither of these fields rests on a solid theoretical foundation? *An accepted full theory of solar activity does*

---

[139] Ibid Landscheidt, T., (1999)

[140] Parentheses by this author

[141] Italics by this author

not yet exist.[142] What we have is only the hope of a future theory. According to P. V. Foukal [143] the mechanism that causes the solar magnetic cycle remains poorly understood, although it has been the focus of intense research during the past half century. There is a lot of literature about *aw*-dynamos, but they are coping with incompatibilities of observation with theory, and they do not offer any explanation of longer solar cycles like the Gleissberg cycle that modulates the amplitudes of the 11-year cycle.

Perhaps, then, the insights leant later on in this book for mechanisms for grand solar episodes might assist in achieving that long-sought after "accepted full theory of solar activity?" In aiming for this a detailed look is taken at what is physically involved in the Sun-earth connection from a top-down perspective. Coincidentally the Gleissberg Cycle is involved, and this will be shown in conjunction with atmospheric manipulation of the Sun in a thin but potentially profitable manner in subsequent chapters.

---

[142]   Ibid
[143]   Foukal, P. V., "The variable sun." *Scientific American*, February 1990, pp 34-41

4. Widening perspectives of our Sun in space: what does the Sun and other star phenomena produce in the Sun-earth climate connection—and what could it do to Earth?

As far as can be seen our entire Solar System is enclosed within what is called the "heliosphere." Indeed, the Sun, Sol, creates the heliosphere. We picture one set of quasi-elliptical hydrodynamic / hydrostatic entities inside of others, like the Russian toy of a round man or woman that go on to reveal another, smaller, round man or woman within; and another within another and so on. The Sun and the entire Solar System all have things "like" Van Allen belts around them and "absorptive" and "deflective" mechanisms that are there due to magnetism: that is, the gas itself, and ionized fields. The heliosphere, then, is a gas-delineated "sack" that encloses other "gas-sacks" made by, and fielded by, celestial bodies and their emitta.

Our Sun of course is somewhere inside the heliosphere as it moves—and it goes thus along through interstellar space. The velocity, always reaching for escape status as in most likely all bodies in celestial motion, is in the Sun's case about 12,500 miles per second. It is aiming in the direction of the constellation Hercules (right ascension 17 hrs, declination + 30). Look up into the summer sky some night and locate Hercules' "keystone" asterism, westward if you can. Our Sun is hurling itself at this constellation at around 12,500 miles per second. As the Sun emits and then strikes the onrushing gas as it moves, a bow shock is formed. This bow shock has great implications for Coronal Mass Ejections (CMEs). In any case it is itself a kind of "magnetosphere" (actually a magnetohydrodynamic sheath). The heliosphere is created to the Sun's front and all around it as it moves in this medium at about the speed just mentioned. This is at the distance where the dynamic pressure of the carried-along solar wind equals that of the interstellar gas pressure. That is the heliosphere in a nut shell (more like teardrop). Our Solar System fits in it, to include Earth and all of us, and every living and non-living thing.

As Sydney Chapman inferred mathematically with Earth—and his delineation can be fitted to the Sun in rough form—the heliosphere is a tear-shaped volume inside the bow shock where the gas's motion is dominated by the solar magnetic fields and by the solar wind farther out from the solar "surface." Its largest diameter, measured at right angles to the direction of motion, varies from about 80 and 100 A.U. In the direction of the heliosphere's motion, it is much larger because of its long tail. Solar coronal regions at high solar latitudes, where magnetic fields are open allow "plasma"—the "solar stuff"—an outward flow. The Sun's magnetic fields close only at a great distance from the Sun.

It could be that the basic hydrodynamic/ hydrostatic description shown in this SOHO diagram (Figure 1) would still be fiction today had not Sydney Chapman and V. C. Ferraro first matched the math to the observations on how bow shocks and cavities form around orbiting bodies regarding magnetospheres. They did this as physicists who thought mathematically, predicting observations. Add magnetism to this concoction and you have what is called magnetohydrodynamics or MHD as regards the Sun, directly, and how one big theory has it operating: the solar dynamo theory.

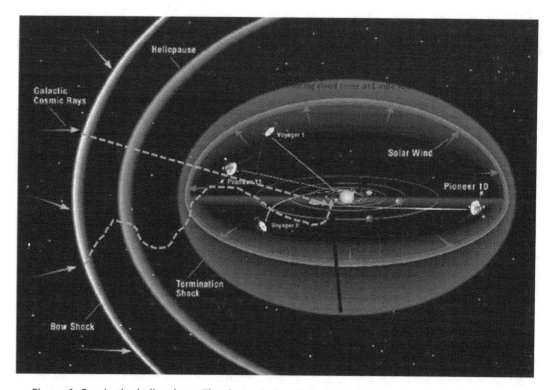

Figure 1. Sun in the heliosphere. The dynamic here is essentially the same as the Earth in the magnetosphere, though they are separate phenomena (see Figure 2). Withstanding shock waves in the heliosphere is experienced by every celestial body in its reach, including Earth. Dashed lines are weak and strong galactic cosmic rays that reach both the Sun and the Earth

What Chapman and Ferraro discerned so long ago was this basic configuration (Figure 2) of how Earth with its magnetic field "fights off" (alternately, "gives in to" from time to time, depending on the Sun's strength) the Sun's magneto-gaseous power—to include particles and the accompanying electromagnetic wavelength fields—from the Sun.

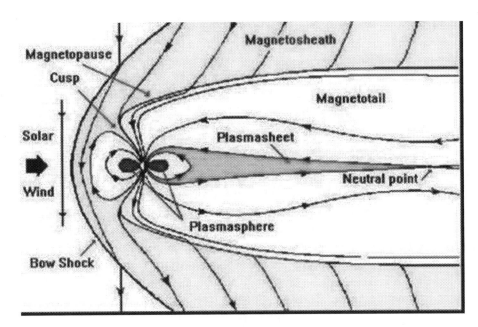

Figure 2. The Earth (that object between the two polar bubble cavities) in its magnetosphere, and its dynamic relation to the "force" of the solar wind emanating from the Sun in its self-created heliosphere. The magnetosphere itself is a sheath of ionized gas.

We look at these two "spheres"—the one of the Sun system and the one of the Earth—systemically. The Earth is encased in this magnetic cloak that forms an invisible "sphere." This is in the Solar System, the Earth taking "it" (the solar magnetohydrodynamic wind) from the Sun.[144] The gas envelope of Earth alternately fights back against it. Some planets, like Venus, have an almost nonexistent magnetosphere [145] but enacts a similar give-and-take struggle with Sol. Yet, abstracting spatially somewhat, we see the Earth (and Mars, Jupiter; all the rest) inside of a bigger bubble called the heliosphere (Figure 1). And, looking at things purely terminologically, like at the magnetopause, the heliosphere of the Sun also has a "pause" (see Figure 1 and compare it to Figure 2: heliopause vs. magnetopause). Both sustain "bow shocks" though not in the same places. The basic resemblance is there. Just as Earth's magnetopause can let in particles from the Sun at the cusp point, there to enter the atmospheric envelope of Earth, so does the Sun's heliopause allow particles in from deeper space. These particles invading the Sun can come from galactic sources; say, from Type 1a supernovae in strong and in weak "waves" (the long dashed lines in Figure 1). These are energetic particles and wind from exploding stars,

---

[144]   Mars and Jupiter, to name two, also have magnetospheres and "take it" from the Sun.

[145]   In addition to weak magnetic sheath (magnetosphere) protection on Venus, the top pole of Venus is almost always pointed at the Sun, constantly allowing for radiative transfers from its closer proximity to the Sun has implications for composition of its atmosphere.

recall. Naturally, the way the Sun and the Earth handles invasive wave/particle blasts is not "just like each other" but in a basic relational sense they handle them similarly.

These particles described in broken or dashed lines in Figure 1 are cosmic rays (to recall, super fast particles that bombard things). Some of these cosmic rays aren't as fast as others (see Figure 1). Their speeds, as with all things, vary. Some, strong, go right past the heliopause, much like some particles from the Sun that aren't as fast at times as at other times, and these go right past the magnetopause, streaming over the magnetotail and over and out into the heliopause. Others, weak, wander into the heliosphere and can, goes the theory [146] of for example, Henrik Svensmark, effect cloud cover production on the Earth one mile up off the Earth's surface (or three kilometers). That is, stronger cosmic rays fired into the Earth's atmospheric envelope could yield higher or more dense Earth cloud cover.

And why do some cosmic rays strike the Sun in its middle at some times more than at others? Cosmic rays do come at us, as the SOHO diagram shows, and yet not all bounce off the heliosphere or the heliopause. If the Sun's heliopause is hydrodynamically weaker for some reason then these rays—coming in from in some cases exploding supernovae within our own Galaxy—will barge right on in though these "pauses." If they do this right on past the Sun's Parker Spiral, or, the Sun's "own magnetosphere" (magnetohydrodynamic sheath) then the chances of these cosmic rays effecting Earth are high, and potentially Earth's cloud cover as well. With a stronger magnetic field the Sun would deflect the rays with a more robust outward hydrodynamic push of gas, particles, and radiation in the Sun's stronger or even "normal" magnetohydrodynamic operating mode. But in cases where this robustness is diminished or has lapsed, in a kind of stasis or equilibrium, the Sun hydrostatically lets things shoot on past it or into it, which in turn means right at us and at times into us (as well as Mars, Venus, etc.).

The usual impact is often nothing dramatic at all (nothing like a large meteor impact for example). This is a different, potentially slower dynamic, invisible to the eye, and hinges on just how much cosmic radiation is getting past the Sun. Or even how much cosmic radiation the Sun itself produces and which our Earth cannot sometimes adequately protect itself from, due to its own tendency toward weakened or looser magnetic fields at times. Earth, then, can be bombarded by cosmic rays from both deep space and the Sun. That the Earth's weather in the troposphere *could* be affected due to more cloud cover created from cosmic rays, the cover blocking the Sun's healing and warming rays, then this in turn would mean *local* climate change for the cooler and degraded weather in the form of more frequently dark and cool days

---

[146]   That cosmic rays can produce cloud condensation nuclei was provisionally proven at CERN on August 24, 2011 in a laboratory. Since satellites etc. do not discern clouds, climate modellers do not see them. It is almost another branch of physics altogether. Refer to Svensmark, H., and Calder, N., *The Chilling Stars: A New Theory of Climate Change* (Totem: 2003); New Experiment to Investigate the Effect of Galactic Cosmic Rays on Clouds and Climate, CERN, PR14.06 October 19, 2006; also, Svensmark et al. Cosmic ray decreases affect atmospheric aerosols and clouds, *Geophysical Research Letters*, Vol. 36 (2009)

for instance; some possibly more windy, possibly more rainy. That Earth cloud cover, howsoever made, can be made brighter once formed and thus somehow be more prone to warm Earth's surface as a theory has been advanced. Long wave radiation beneath clouds can hem in Infra Red (IR) and cause fantastic short term and local warming. But this rules out convection forces conducting energy transfers underneath these very clouds, convection breaking up the IR concentration.

Additionally when the Sun is weaker, it is less bright: something the distance between the Sun and Venus, for instance, denies far the less. Earth is further from the Sun than Venus. In solar weaker times then, unevenly across the hemisphere, we get more cloudy days in seasons or in places where clear high pressure weather often should predominate, the actions of the aforementioned Hadley Cell notwithstanding. Solar-emitted long wave radiation for instance off cloud cover from above reflects "warmth" back out into space—which in turn, causes more cloud-propagated weather underneath it as a "feedback"—the positive or negative label to feedback depending mostly on personal perspective. This is the so-called albedo effect as discussed in just one of its incarnations regarding cloud cover.[147]

Earth's upper atmosphere is studied by scientists as the field of aerology. We look at the navigational chart-mapped zone called the "North Atlantic high pressure system" used in practical navigation. Due to the practical concerns involved the nature of these systems is often given a carved-in-stone quality that betrays their very variable behaviors. According to orographical [148] or other aeronautical/meteorological-to-terrain maps for professional fliers and sailors they show air pressure gradients in the "usual" areas, located along the eastern US seaboard and somewhat out to sea of that location. This long sausage-shaped area is sort of a "constant high pressure system." It is much like the similarly-charted "Greenland low pressure system" as being a kind of "constant *low* pressure system." Such are the opposite dynamics. In fact clear weather often dominates the former and cloudy, the latter. Or, put in a more aerological way, you could say that high pressure systems (non-cyclonic) dominate the one, with much long-wave radiation falling to Earth's surface (effects of an unobstructed Sun): and low pressure (cyclonic) dominates the other with a concurrent solar-blocking state that is normal for that "constant" low-pressure gradient zone. But the "mean" is usually high pressure in the one, versus low pressure in the other. (This is precisely why the distinction is made: for mundane knowledge of what to expect from the one zone as you fly or sail through it versus the other.) If more cosmic rays are creating more cloud condensation nuclei in the lower atmosphere generally, then the "constant" North Atlantic high pressure system may be prone to more low pressure events, in spite of the usual terrain (orographical) and other

---

147   Another is sunlight reflecting off Earth when rain forest canopy has been removed; large snowfields and glaciers, and even water. Yet another is shiny volcanic dust in stratovolcanic clouds that additionally can mirror away the benign light of the Sun.

148   Orography refers to mountain studies; in this context, regarding how weather patterns are influenced by the terrain in the lithosphere they occupy within or around weather patterns. and vice versa.

considerations keeping or holding toward the usual expected "mean." The kilopascal averages "typical" in one zone are suddenly all out of typical alignment, since the pressure has altered on a microclimatic level; the microclimate being altered by the effects of cloud condensation nuclei (that is, more clouds, with storms).

The effects of the previously-discussed "850 B.C. event" Hadley Cell circulation due to a weaker sun causing "storm tracks to move more equatorward" quite possibly has something to do with pushing these otherwise-predictable "pressure system zones" out of the "normal" mean. These patterns like the this or that low-pressure system are variable and prone to being thrown out of kilter. It is emphasized here that this changing is happening on a *local* level, from the usual perspective of global levels which also of course occur. Albedo effect like reflection of shortwave radiation off the top of cloud cover (and long wave IR under it, affecting Earth surface temperature in short terms) aids in propagating yet more cloud tracks and low fronts in either "normal" zone (North Atlantic or Greenland) depending on how "local" the effects of Earth surface reflectivity is spread out across the hemisphere. (Refer back to the discussion "Powerful regional effects of climate change vs. 'global warming': three views" in Chapter 3.) In such cases, the "Greenland low pressure system" may be more prone to cloudy weather and other events involving low pressure (like storms) than is usual. And the "North Atlantic high pressure system" may experience more low pressure ones as well. Even more confusing, the "typical" pressure gradients might be reversed at times, so that the "Greenland low pressure system" experiences more high pressure events, whilst the "North Atlantic high pressure system" goes though more low pressure ones. Such is the state of our understanding. These two, the North Atlantic High and the Greenland Low, are just two examples in this consideration. Aerological and orographic maps delineate many more. Thus, like the Sun (inconstant as it is) the atmosphere too is prone to inconsistency from solar particle blasts (just one factor that influences it, of course) howsoever much we impress consistency in some patterns so as to obtain an orderly view in very practical matters such as air, sea, and space navigation.

That is the condition on Earth, and the atmosphere's condition is somehow handled by the Sun. The Sun itself, then, is a gaseous and unstable boundary condition-shocked system, as is to a degree the heliopause, though the two are not the same. The Sun creates the latter. The latter may help shape the Earth's atmosphere and its climate. As meteors can come from deep space and penetrate our atmosphere so can rays of a "cosmic" particle variety. These particles may affect Earth climate, the "coupling" of chemical particle ingredients that does this almost always happening in Earth's stratosphere (or higher). Such complex coupling, of which still is little known, occurs in what is labelled the "plasmasphere" (dark bubbles shown in Figure 2). That the effects of coupled particles is allowed to drift down into Earth's atmosphere by a weaker Sun, alternately "ripped away" (into the "plasmasheet" see Figure 2) by a stronger Sun, gives us a starting point as to why theories of natural Earth climate alteration exist from such causes in the first place.

Casting aside the groping for relationships on which to build meaning, and considering the odd and strange nature of most of what was just considered, the matter of Galactic-Sun-earth climate connecting is infinitely complicated and shrouded in doubt. It is even looked on with mistrust in science. There is even outright disbelief. This is the case within practical and theoretical science and especially out of it. Institutional science must often proceed in tiny boxes, separate, and buried tomblike in specialized jargon, creeping at a snail's pace. Some convergent connections are made. But being un-testable at times these convergences are justifiably suspect. Something is there for all that. These alterations of our climate such as they would occur are entirely natural, hardly as capricious as it is sometimes made out, and have their own evolved and natural checks keeping us quite safe in a series of trade offs, or, total or near-to-total damage-negating events. Our own consistent human existence for over 50,000 years is proof that some or many natural processes conspire to keep us here in that trade-off's give and take—though we can always learn more to understand and overcome finer-point obstacles to enhance our well being.

The Sun is a gaseous. unstable, open-boundary system that has more than "plasma" emanating from it and headed toward Earth (despite the labels of plasmasheath, etc). There is unpredictable magnetized particle ejecta mainly from solar flares and are sometimes called Solar Energized Particles, or "SEPS," coming from these flares. SEPS are particles, versus, say, plasma clouds from CMEs, SEPS being the manifesation of a hydrodynamic structure. (Plasma clouds are also ejected from coronal holes and are "broadcast" by the solar wind). SEPs versus plasma ejecta are semi-relativistic particles with velocities between a few tenths of the velocity of light or very close to light's velocity. They penetrate to a considerable extent Earth's magnetosphere and outer ionospheric layer. How much they penetrate the Earth's atmosphere (as relativistic implies) depends, naturally, on the various SEPs' energies. Separately (though no less importantly) the Sun also just as unpredictably bathes us in the entire contents of the electromagnetic spectrum (though not evenly so): most noticeably for our ensuing discussion, Gamma rays (Γ), some Infra Red (IR), much Ultraviolet (UV.) and the high-temperature, mostly solar-wind bearing X-rays.

This is an interesting give and take, then: a kind of *dynamic* that could at least be assisting climate modulation in the Northern Hemisphere (we recall the discussion on cosmic rays).

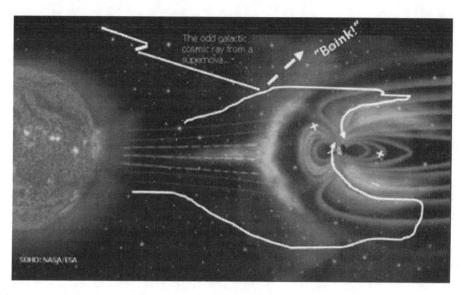

Sun-earth cartoon showing ejecta from the Sun striking Earth's magnetosphere. The Sun in this cartoon is quite strongly emitting, but the Earth's magnetic envelope is <u>also strong,</u> much solar ejecta streaming past. It shows some seepage, however, of magnetized particles ("ejecta") going into Earth's "plasmasphere." An odd galactic cosmic ray, probably from a supernova (exploding star) with low surface brightness (and so, unseen) weakly bounces off the Sun's and Earth's protective sheaths.

Case one: weakened solar state ("cooler northern hemisphere")

- The interference of cosmic rays heading past weak solar and Earth magnetosphere for making Earth cloud cover, beginning the chain of events that could mean cloudier weather at unpredicted times, with attendant increased (cold) precipitation, and the storage of Greenhouse Gases (to include Carbon Dioxide). Carbon Dioxide is not absorbing IR radiation (and at such times there is most likely less IR in concentration).
- Together with a weakened solar magnetosphere and presumably reduced solar wind, more SEPS when released by solar flares (and the hydrodynamic nature of plasma, from CMEs, out of coronal holes) enter Earth's atmospheric envelope (which deep time proxy records in snow, etc., in abundance reveal)—even if SEPS, when emitted, are sent in all directions around the Sun and heliosphere at ejection but at very varying rates (open boundary, unstable, and non-linear). Less sunspots are often an attendant symptom of decreased overall "Total Solar Irradiance" (TSI). The Sun is *less* bright.
- Together with cosmic rays, and a weakened solar magnetosphere and presumably reduced solar wind, the Sun is overall less "bright" . . . and it is emitting lesser/weakly, or both, in all directions around the Sun and heliosphere constantly but at very varying rates [149] (open boundary, unstable, and non-linear)—which means less Γ rays, UV, X-rays,

---

[149] The word is differential.

and possibly IR. (Agnes Clerke's "reduced aurora" and other limited light emissions, along with a "profound magnetic calm.") Earth might be more prone to CME/solar flaring effects if they are "dead on" strikes, as Earth's magnetic sheath is weaker, overall, thus not blocking the strikes so well.

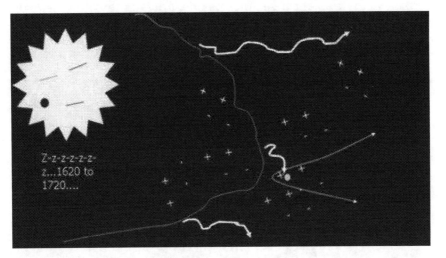

Weak magnetic state: "profound magnetic calm" . . . even weak grand phase (episode)

Case two: strong solar state ("warmer northern hemisphere")

- The reduced interference of any cosmic rays past strong solar and Earth magnetospheres could reduce excessive cloud cover, beginning the chain of events that mean warmer, drier weather at unpredicted times, (warm) precipitation, with attendant Greenhouse Gas release, like Carbon Dioxide, which absorbs IR very well and so increases or magnifies the (mostly *local*) temperatures in microclimates. Cosmic rays are mostly deflected by the Earth's magnetosphere.

- Together with a strengthened solar magnetosphere and presumably increased solar wind, SEPs when released by solar flares (and the hydrodynamic nature of plasma, from CMEs, out of coronal holes) may bounce off of the magnetosphere or get into Earth's atmospheric envelope (which a lack of them in deep time proxy records in snow, etc., reveal)—even if SEPS are sent in all directions around the Sun and heliosphere at ejection but at very varying rates (open boundary, unstable, and non-linear). Most of these SEPs, ionized, highly-charged, zip past Earth out through the "plasmasheath," and back into the wider heliosphere. More sunspots are often an attendant symptom of increased TSI. The Sun is *brighter*.

- Together with less cosmic ray interference and a stronger solar magnetosphere and presumably increased (and so, hotter) solar wind, the Sun is overall "brighter" . . . and it is emitting more/strongly. or both, in all directions around the Sun and heliosphere constantly but at very varying rates (open boundary, unstable, and non-linear)—which means more Γ, UV, X-rays, and any IR. It would have been the opposite of what Clerke and Maunder noted in the records from England in the early 1600s to early 1700s. That is, a brighter Sun would have caused more auroral events up and off the Earth's poles witnessed perhaps more often by Viking explorers in "Vinland." Earth might conceivably in these solar active times be under greater threat of "dead on" CME/solar flaring effects. But an increased and potent Earth magnetosheath that occurs naturally in such times would guarantee added protection from them.

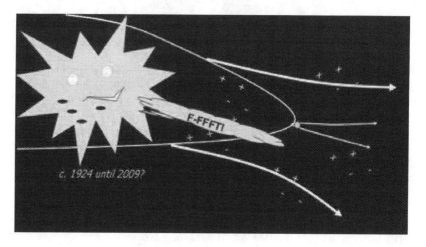

Strong magnetic state: "magnetic hyperactivity" . . . even *strong* grand phases

In the cases of the usual give and take of "regular" solar maxima and minima, these dynamics, as bulleted above, for the "cooler" or the "warmer," are not so pronounced. They can be more "regular" and not so extreme. On the other hand, if markedly pronounced, they could be signs of "grand" solar extremes in either the weak or the strong dynamic. What "extreme" or "normal" constitutes in terms of microclimate weather leaves much to interpret if all or any of it indeed does effect weather in the long or short term.

◊

It is quite hard to explain or get context on this without animation: that is, be able to explain and to draw the 3D-like proportions of these complex "coupling processes" between the stellar body of the Sun and the planets it "rules"—Earth being one of them in this regard.

The Sun shoots out its SEPs mostly from solar flares, additionally magnetized plasma from CMEs, along with any spectral wavelength emissions (Γ, UV, X-ray . . . IR) nonlinearly and independently of each other (sometimes on, sometimes off, sometimes in different directions, at different rates and different times; sometimes at the same time etc.). [150] The emission of wavelengths is also recondite and likewise as "differential." How much the Earth's atmosphere draws in of SEPs, magnetized plasma from CMEs, and wavelength emissions depends upon a lot of factors. Earth's orbit is one such factor, as shall be discussed.

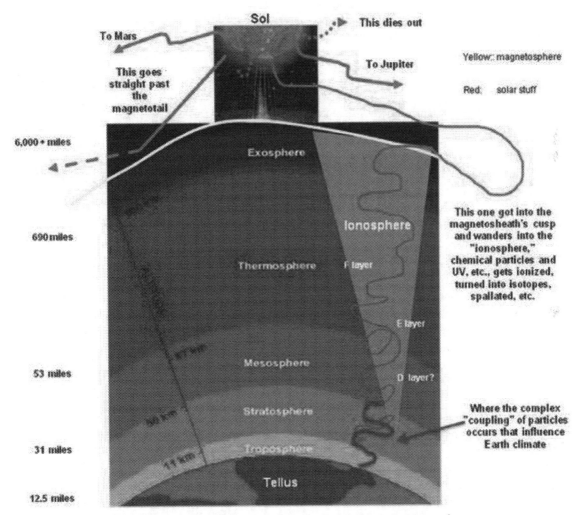

Cartoon of atmospheric layers abutting the Sun to Earth's surface

Starting at the very top, looking down, the Sun "owns" the Sun-earth exosphere and it is here where the Sun's SEPs (but galactic cosmic rays too) may make their ways, bullet-like, into the cusp of the Earth's magnetosheath, to whip around and then some on down into

---

[150]   The word, again, for all this, is differential.

the "plasmasphere" and into Earth's lower atmosphere . . . particularly one area of it that floats around called the ionosphere (the long triangle pointing downward in the cartoon on atmospheric layers). Here is where they might fail as well and wind up spun back out into the heliosphere. It is mainly open space over 6,000 miles from the Earth's surface; it consists of the same stuff the Sun is made of (mostly Hydrogen). To generalize, you could say there are almost no particles up here at all. The few particles present are presumably miles and more apart and from each other. The particle density in the lower atmosphere dwarfs the particle amount and density found in the upper atmosphere.

The interesting thing about the next layer down, the thermosphere (at the top almost 700 miles from Earth's surface) is that it struggles with the Sun's activity, it being pummelled freely by the Sun's excessive action or more or less ignored and let loose to expand when the Sun is weakly variable. As such, it richly illustrates the dynamic push on the Earth as shown in Figure 2, right where the bar showing "solar wind" appears in front of the diagram of the Earth's magnetosphere. So the "height" of this level of "atmosphere" from the Earth's surface, importantly, is also variable. Like the Sun. Depending on how strong the hydrodynamic solar wind is in the solar corona, the distance between it and the thermosphere "ring" may be many Earth radii (that is, several diameters' lengths of the Earth)—if that wind is more hydrostatic and thus, flaccid or weaker. Or, if that hydrodynamic push, carrying "solar stuff" with it (SEPs, plasma, etc.) strongly at Earth is very strong then that magnetospheric "ring" may be only one or two Earth radii thick and the magnetosheath be pushed straight back in the demonstration of bow shocking shown earlier (Case One and Case Two). Thus, the Earth's atmosphere is also very strongly controlled by the Sun at this height—and, at certain levels of the ionosphere, downward, into the lower atmosphere, does the Sun control particle blasts. The temperature in the thermosphere is weird: it *increases* with height up from the next layer down (mesosphere) and, as the layer expands and contracts with highly variable strong and weak solar activity, conditionally it can be said that the temperature here is also affected by the Sun. It can reach 1770 Kelvin (over 2,000 degrees °F). Not a cool place.

Beginning in perhaps the exosphere at times, and at other times, lower in the thermosphere, is found the ionosphere. This "layer" may be due to the very UV radiation emitted by the Sun itself. So one entire "layer" of Earth's atmosphere is possibly the creation of one band of solar wavelength emission. And like an allegory of the Sun's activity, the ionosphere winnows its way down all the way through all the other Earth atmospheric layers to the troposphere. The ionosphere is a "layer of electrons and electrically charged atoms and molecules that surrounds the Earth" stretching from a height of about 31 miles to more than c. 700 miles, and most likely varies due to the Sun's strength. When it touches the stratosphere, it is where the "indirect coupling" of the Sun's particles occur, and which probably affect lower Earth climate directly, in the troposphere and thence lower to the very surface of the Earth. Exactly how this magnetized dance occurs and what it does to the Earth's climate (to include weather if at all) is unknown. It is an intense object of study by NASA's Solar Dynamics Observatory EVE subsystem.

Aurora borealis spreads here at the lower portion (of the mesosphere) seen mainly over the poles, from CME-induced particles compressed in the magnetosphere. If the Sun's activity is strong enough, aurorae will creep down into the lower Earth latitudes. In the mesosphere, the height from ground zero on Earth is about 53 miles. There is more density and friction at this point: where old re-entry vehicles like returning Moon capsules would superheat, and where meteors burn up, and where sometimes, being too large, they do not do so but strike Earth. This is a protective zone for Earth indeed, if it fizzles these things out, and as a natural force, it mostly does. If it wasn't for this natural belt of protection we would not be here. A meteor large enough to pass through this layer would not burn all the way however, which is why we sometimes see bolides burning up in the lower atmosphere, having been significantly reduced in size in their rides downward from places like the Kuiper Belt on the outskirts of our Solar System. The puncturing of this layer without burn-off of meteors / asteroids a quarter mile in diameter and up have purportedly destablized Earth's climate in the past.[151] Unlike with the thermosphere above it, the temperature *decreases* with height in the mesosphere. It is a cool place (all the way down to −121 °F, or 188 Kelvin). It is the coldest "part" of Earth. Noctilucent clouds form here since gases in them condense, and these are seen from Earth's surface.

In the stratosphere ("ceiling" at 31 miles) the temperature also increases with height, which is the exact opposite of the troposphere below it (capping at 12.5 miles) all the way to six miles, below which we have as our common breathing space. Density increases rapidly from here all the way down. By the time it reaches sea level the particle content is immense, to include things like Earth's dust, salt sea crystals, ice crystals studded with anthropomorphic and natural air pollutants (hydrocarbons, sulphur aerosols) pollen, and at micron levels and smaller, traces of "solar stuff's" (CME and solar wind plasma's) isotopic particles like Beryllium-10 and Carbon-14 and many, many more, all in the process of being connected to other particles being free floating or absorbed by biomass (trees, snow fields, attached or detached glaciers, etc.).

The "middle atmosphere," or "electrosphere"

When you take the mesosphere, stratosphere, and troposphere together, one finds what is sometimes called the "middle atmosphere" (especially the mesosphere and stratosphere) and, from ground zero to about 50 miles up, you have what has even been called by experts the "electrosphere." [152] Electrical conductivity from these locations downward varies, and with the density involved in the lower levels, aerosol particles are most prevalent. "Measurements of V/m fields with symmetrical and redundant sensors appear to be real in the mesosphere. These fields complicate the `mapping´ picture of electrical coupling and may also modulate the transport of

---

[151] Such as the Cretaceous-Tertiary Extinction Event ("K-T" event)

[152] Herman, J.R., and Goldberg, R.A., *Sun, Weather, and Climate* (NASA: 1978)

aerosol particles" [153]: possibly even cloud condensation nuclei [154] that is modulated by cosmic rays reaching this area much farther down from the mesosphere due to ionospheric activity. Hence we have storms, electrical, thunder and lightning, and so on in the thicker atmosphere that stands to receive the many ionized particles here, "way down under." Thunderstorms are partly the result of cloud propagation from the earlier-described nuclei. This "layer of electrons and electrically charged atoms and molecules that surrounds the Earth," then, cuts all the way down through the layers of Earth's atmosphere like a knife. And this very "layer of electrons" is itself controlled by the Sun.

One of the most important parts of solar radiation in these transfers in the atmosphere is ionospheric UV, but also X-rays, particularly when the Sun is very active, as X-rays come most to only off of the hot, and so, very solar-active solar corona. As the Sun becomes more active, energy transfer increases. Atomic particles moving within the electromagnetic wavelength bands are energized enough to shoot photons (light blasts) popping electrons from their shells (at a subatomic level) as they go from one series state [155] to another. This for example as they might through the ionospheric D Layer (refer to the cartoon on atmospheric layers). There is a high velocity involved and, in the transfer, a rise in temperature. This explains why, in areas of massive ionization in the atmosphere [156] there are odd areas of extremely high temperatures (like in the thermosphere) sandwiched between areas where electrons are captured by positively charged ions, reversing the ionization process, and this is done in cooler zones as the particles drift downward. The lower and lowest parts of the atmosphere being the densest of all (all that dust and sea salt crystals etc.) the act of *recombining* happens most of the time. This is one reason why you find less severe temperature rises, more particles, and what is termed weather events (thunderstorms, rainstorms, wind storms, tornadoes, and so forth). For "recombination" is the opposite of "ionization." A little-known fact is that Earth can itself produce high energy rays, such as Gamma (Γ) for example during extremely violent thunderstorms—and hurl these out into space. Earth itself thus contributes to the cosmic muddle in the heliosphere.

That area of the stratosphere where the anomalous "coupling" so often spoken of by solar and geo scientists occurs is in the D Layer of the ionosphere's cheese-slice spin into the stratosphere. Here, X-ray radiation from times when the Sun has a very active solar corona ionizes nitric acid, [157] which puts the boom into thunder much lower down, as this is most definitely part of the "electrosphere" in our troposphere. Recombination is higher and ionization, a lot lower. The cosmic rays here cause disturbances in radio frequency on Earth dayside (AM radio is

---

[153]   Hale, L.C., "Middle atmosphere electrical structure, dynamics and coupling," *Advances in Space Research*, Vol. 4, No. 4, 1984, pp 175-186

[154]   Still very debated

[155]   As in Lyman and Balmer Series

[156]   Dependent variably on the Sun's activity for the greater, the season—say, high summer, and time of day—say, mid morning etc.

[157]   Nitre is made from this, which is an active ingredient in explosives.

prone to such a weaknesses in the daytime). During so-called "solar proton events," (another word for SEPs) they can wash out radio signalling—and the satellite-to-earth signalling of low orbiting satellites that deliver say television can be particularly harshly hit when solar activity is nonlinearly high. If the SEPs (or, proton events) come from a particularly strong solar flare/ CME bow-shocking event, the potential for Earth infrastructure destruction is there (as described regarding Carrington's 1859 observation in Chapter 2). Yet if not protected from "deep-space" cosmic rays by a stronger heliosphere (stronger, and so hotter, Sun) we are "damned if we do and damned if we don't" since we no longer have closer solar cosmic rays to worry about. We may at times have unpredictable stronger galactic cosmic rays interacting with the contents of the ionosphere/electrosphere due to a weaker heliosphere. At such times, we can only hope that we do not have too many stronger galactic cosmic rays from exploding stars (supernovae) penetrating Earth's magnetosheath. This "because the primary source of ions in the . . . troposphere is Galactic Cosmic Rays (GCRs), their role in atmospheric nucleation is of considerable interest as a possible physical mechanism for climate variability caused by the Sun."[158] For all this concern, the natural give-and-take of Earth in the Solar System's checks and balances must be recalled in order to maintain calm.

How to detect GCRs in advance in times of solar weak periods is a matter of concern at NASA [159] if only for satellite and manned spacecraft early warning protection, for example. But there is also the climate change question. Experiments have recently shown [160] that the suspected culprit behind cloud condensation nucleation in the troposphere (but possibly not in the non-boundary areas) is the ionization of the Sulphuric Acid-water (H_2SO_4-H_2O) and Ammonia (NH_3), a compound of nitrogen and hydrogen. Together this is called "ternary interaction of gases." Perhaps increased GCRs into the Earth climate system recently have aided in cloud formation, globally? This allowance of GCRs into the atmospheric system is due to weakened magnetic fields of the Sun and Earth—the heliosphere (see Figure 1)—due in part to reduced solar activity. That is, the GCRs come crashing on it since they are not repulsed. But even the authors of this paper [161] admit:

1. Quantitative measurements of the roles of ions and ternary vapors are lacking
2. The nucleation mechanism and the molecular composition of the critical nucleus have not been directly measured
3. It is an open question whether laboratory measurements are able to reproduce atmospheric observations: recent experiments have concluded that atmospheric

---

[158]   Kirkby, J., et al "Role of sulphuric acid, ammonia and galactic cosmic rays in atmospheric aerosol nucleation," *Nature*, Vol. 476, 25 August 2011, p. 429

[159]   In 2010 NASA reported the highest amount of cosmic rays in recent years since record keeping for it began.

[160]   Ibid, Kirkby *et al*

[161]   Ibid, Kirkby *et al*, p. 429

concentrations of H_2SO_4-H_2O without ternary vapors are sufficient [162] or insufficient [163] to explain boundary-layer nucleation rates.

The Kirkby *et al* experiment at CERN in August 2011 confirmed parts of Henrik Svensmark's theory on cloud condensation nuclei being created by cosmic rays (which we recall are not "rays" at all but energy bundles.)

This was the long-delayed and awaited CLOUD [164] experiment. This was a climate chamber; a c. 120 inch-diameter stainless steel canister containing humid pure air, sulphuric acid (SO_2) ozone (O_3) and ammonia NH_3 gas. Charged pion beams (that is, charged particle beams) were allowed to strike the canister. The behavior of the particles was then watched at various temperature levels. The cosmic rays changed the aerosol creation rate over a factor of 10. Does this linkage show cloud condensation nuclei actually making clouds such as nimbus or stratus in the sky, which are great reflectors of long wave radiation and so, deflect warming rays (etc.) off the Earth? No.

But although this is not watching it happen in actual Nature, the laboratory was located in the troposphere, three miles on average lower than where such cloud condensation nuclei would actually form and as such, is a far better demonstration of the same than any done on computer models.

---

[162]   Sipila, M. *et al.* "The role of sulfuric acid in atmospheric nucleation," *Science* 327, 1243-1246 (2010)

[163]   Metzger, A. *et al.* "Evidence for the role of organics in aerosol particle formation under atmospheric conditions," *Proc. Natl Acad. Sci.* USA 107, 6646-6651 (2010)

[164]   Cosmics Leaving Outdoor Droplets (CLOUD)

(Left to right). "Particle coupling" illustrated. A molecular cluster of gases such as those used in the CLOUD experiment in August 2011 would, in Nature, presumably due to nucleation and evaporation in the "electrosphere" (see D Layer in the cartoon of atmospheric layers) have a charged particle cluster that eventually forms via organic interexchange yielding an aerosol particle. This particle is further degraded organically until it becomes a cloud droplet nucleus (large ball, far right). These in turn could form extra cloud cover far down on Earth in times of weakened solar activity, since GCRs have freer reign in the weakened magnetosphere of Earth. (Courtesy of CERN)

Nonlinearity versus linearity: a matter of concern in Sun-earth climate connection understanding

The above-information could best be studied by those atmospheric scientists who are looking for more clues to climate change as modulated by the Sun.

Yet a persistent major bone of scholastic and institutional science contention as to the non-interference in the Sun in "global climate change" is based on some valid assumptions. Again, we return to the lack of evidence for short-term global super-warming by the Sun (the as-yet still anomalous long-term effect noted via proxy data being admitted as a *fait accompli*). These discordant chimes are trilling toward concordance.

The Sun, some researchers point out, just cannot be heating us up in a major way in the short term (or not cooling us down, either, if the above information on the GCRs bombarding us die to a weaker sun is rejected). No direct evidence for it from proxy data, and unclear understanding of the solar chromosphere and how it interfaces with IR and UV, leaves the question of shorter term high-level warming, whether local or global or otherwise, muddied.

Thus we come to a valid topical and much-debated question of the cause of the strong terrestrial heating in the last few decades of the 20th Century and to the very first few years of the Twenty First. Much of it had been argued as being human-made. Outside of "how much is that exactly from human sources vs. natural ones," the assumption is not trivial but as good as any. Are we contributing so much greenhouse gases from our Northern Hemispheric industrial activity since at least 1940 and its residual effects to short circuit the global climate? The evidence for us being responsible for local warming around industrial and populated areas is very much in evidence and indeed is measureable. The noted "for the warm" climate change over the previous 20 years is usually attributed to greenhouse gases whatever the source (natural or man-made). There *is* a greenhouse effect on Earth: Carbon Dioxide is pinpointed as a major source of the warming (and not water vapor—another "greenhouse gas.") [165], [166]

But findings show that the warming of the past few decades is restricted to industrialised areas (de Laat and Maurellis, 2004 [167]) *locally.* There also is an observed absence of significant global heating hemispherically, shown by satellite data. Thirdly, there was a marked "constancy" of

---

[165]   I point out evidence for data-tweaking from authorities such as IPCC author Richard Lindzen (most recently in US Senate testimony in 2010) and the revealed data manipulation at the University of East Anglia and Penn State.

[166] CO_2 is strange, most likely due to the Carbon element. In the closed-system experiment Biosphere II, an oxygen decline happened but a corresponding increase in carbon dioxide did not appear. An investigation by Jeff Severinghaus and Wallace Broecker of Columbia University's Lamont Doherty Earth Observatory using isotopic analysis showed that carbon dioxide *was reacting with exposed concrete inside Biosphere II to form calcium carbonate, thereby sequestering the carbon dioxide and, as part of it, the oxygen that had disappeared.* (Severinghaus, J.P., W. Broecker, W. Dempster, T. MacCallum, and M. Wahlen. 1994. Oxygen Loss in Biosphere 2. EOS, *Transactions of the American Geophysical Union,* Vol. 75, No. 3, pp 33; 35-37). In an open boundary system like the planet Mars, a radar investigation recently showed 12,000 km of frozen CO_2 under the planet's surface, "about as much as is in its atmosphere today. This suggests that the thickness of the Martian atmosphere may change several fold over the roughly 100,000-year warm and cold cycles that are caused by . . . axial tilt and orbital eccentricity." News Notes, *Sky & Telescope,* July, 2011 Vol. 122, No. 1, p. 17, The same thing may just happen to Earth's reservoir of CO_2 over periods spanning the Milankovitch Cycle (c. 100, 000 years).

[167]   De Laat, A. T. J. and Maurellis, A. N., "Evidence of Influence of Anthropogenic Surface Processes on Lower Tropospheric and Surface Temperatures," *Int. J. Climatol.* 26: 897-913 (2006), (online) 27 March 2006. "Our analysis of climate model simulations of [Greenhouse Gas] GHG warming confirms our earlier results (Paper I), namely, that they do not show any kind of CO_2 emission—temperature trend correlation. In fact,

temperature between 1998 and 2008 hemispherically (it was not increasing between these times). We look then at the Sun once again for climate variation.

"There is an observable solar signal in the troposphere; it depends on latitude, longitude and height in the atmosphere." What does this mean? It can be said that in some cases [168] that this statement can be modified to add quarter or half-century (multidecadal) and century (centennial) footprints in the solar signalling as regards a direct solar impact to at least a part of the Northern Hemisphere's climate, depending on "latitude, longitude, and atmospheric height" and the solar-input of Total Solar Irradiance (TSI). Testable results of the Sun's direct modulation of climate over the Arctic in multiple-decade and century-level timescales is threshed out.[169] There was a wish to connect more decade-present solar climate variation to the more "stable" thousand-year evidence for the same as shown by Braun (2005) Weng (2005) and Dima and Lohmann (2009) in that paper. It was a functional attempt to tie observation of solar influence on climate locally in a shorter timeframe. The keyword here is "local" or "regional"—as was discussed in detail in Chapter 3.

The physical investigation of the origin of the Sun-earth climate relationship (Earth temperatures) based on a unique delta (Δ) time-curve, assumed to be valid for the whole Earth's surface, is basically untenable for describing how Nature operates. For climate modelling, "delta T" indeed has a specific use. For building bridges of understanding as to what may be going on, such time-curves have uses. But the delta averaged temperature of Earth is often upheld as the ultimate truth of our understanding Earth climate temperatures and Nature. This is like taking a clean color textbook picture of a truck and then assuming that this is an actual moving, say, fifteen-foot long, tire-on-road grinding, fuel and oil burning truck.[170] Though the holistic approach to problem solving involving ideal entities, perfectly defined, can give an exhaustive if thorough understanding of a science problem, it does little for the focus of the actual solving of specific problems (recall Landscheidt's view on the lack of a unified solar theory and his insistence that "we cannot know it all" in Chapter 3). Loci must be identified and separated from the totality—whilst admitting to that very totality. There are specific questions to be asked about Nature to divine a partial understanding that is useful and knowable. That is how humans first became successful at mastering aspects of Nature in wide scale manners. Isaac

---

the modelled temperature trends are quite insensitive to the magnitude of the industrial CO2 emissions." (p. 909)

[168]   Soon, W., "Solar Arctic-mediated Climate Variation on Multidecadal to Centennial Timescales: Empirical Evidence, Mechanistic Explanation, and Testable Consequences," *Physical Geography*, 2009, 30

[169]   Ibid, Soon (2009)

[170]   As Plato was purported to have said, that there was the actual table, and then the picture of a table, then, we have his two separate "realities." Some prefer textual realities: others admit to many realities beyond their control (textbooks being ultimately "controllable"). Attempting to control large sections of Nature is impossible. Attempting understanding for shots at control of some sections of Nature, piece-wise, is practicable.

Newton had such trouble in explaining "the corporeity of light" when enquirers demanded he define the problem entirely in an hypothesis:

It is true, that from my theory I argue the corporeity of light, but I do it without any absolute positiveness, as the word perhaps intimates, and make at most a very plausible consequence of the doctrine, and not a fundamental proposition . . . had I intended any such hypothesis, I should somewhere have explained it. But I knew that the properties, which I declared of light, were in some measure capable of being explicated not only by that, but by many other mechanical hypotheses; and therefore I chose to decline them all, and speak of light in general terms, considering abstractedly as something or other propagated every way in straight lines from luminous bodies, without determining what the thing is. (Horsely, Newton: 1779-85) [171]

As we know, he was correct in finding out some of properties of light he investigated—even if he could not give an exhaustive hypothesis that explained all aspects around it in a neo-platonic epistolary manner. Note also that he used, as did Kelvin and other early dynamicists, the notion of "all things moving at once, equally" in the specifics of his observations. This is metaphysically very limiting. However it does discover what parts of a thing are, being examined: much as a any mechanic must do an empirical check of whatever they are tinkering with. Due to Newton's inquiries on such a high plane in so low a manner of checking, useful knowledge on optics broke forth unto the world as if a great dam of ignorance had been shattered. Other investigators used it later on. Similarly do we try to explain the phenomena on the Sun without worrying about knowing "what the (whole) thing is." It is not necessary and we cannot "know it all" as Landscheidt might have said. We can know a lot about something without even knowing (in a detailed scholastic sense) what the thing really, absolutely, is. Unlike with Newton and Kelvin, however, any potential connection the Earth's climate from the Sun as moving SEPs, CME plasma ejecta, and wavelength emissions is—its complete "irradiance"—the total Earth temperature cannot be correctly measured, even if parts of it can be, locally, due, in part, to height, latitude, etc, as Duhau and de Jager suggest and like one rich study shows.[172] Here, then, the aim is not to look at "all things moving at once, equally" but to refine this into the details of locality, specific measurements, and to take into account the many diverging and

---

[171]   See Anstey, P.R., "The methodological origins of Newton's queries," *Studies in the History and Philosophy of Science*, Vol. 35A, 2 June 2004 pp 247-269. It is clear Newton wished to set a strong precedent in the proceedings of experimental natural philosophy from his time, hopefully onward. Words and lengthy definitions, craving an impossible exactness not necessarily useful even if exact - and the inevitable "disputes" they brought about - could open up questions that lead one away from an up-front, usable fact or set of consequences drawn from observed Nature. It can lead the "natural" philosopher astray from the sanctity of the observation itself if it is not workable in some experiment. And many observations, noted Newton, could not be worked into tidy experiments. Perhaps most importantly, if incorrect verbalisms lie about, pristine in logic but barren for the most part of hard proof, how much would such encumbrances someday stymie future natural philosophers?

[172]   Ibid, Soon (2009). Also, relevant papers by Duhau and de Jager

different movements of the phenomena all at the same time, whenever and wherever possible, in a moving, useable dynamic.

Looking ahead to the next chapter, what could be compounding this reliance on the admirable if ineffective "delta T" (time curve) temperature measurement on Earth in geophysics is perhaps an over reliance of astrophysicists on thinking that if there is a *linear* relation solely on the Sun for sunspots = irradiance / maxima = some climate effects on Earth, then there must be a *linear* climate answer in "delta T" on Earth, reflecting backwards and up to the Sun. That is, there is an equally linear relation between climate, globally, and "global temperature," tied neatly to the Sun, and the best of which is but a weak and inaccurate average that shows no temperature specifics locally (regionally) that would tell a far different climate story.

If TSI—("the cosmic flux") is calculated along with sunspot groups, it is assumed that all variables associated with solar activity linearly depends on the sunspot number.[173] Yet, TSI and sunspot groups bear a nonlinear time relationship with the Sun's toroidal (maxima-making) magnetic field, of which sunspot number is a proxy. That is, the sunspot number is but proxy of the solar toroidal magnetic field. TSI and sunspot group number should not be calculated together. Sunspot number says little of the direct magnetic power emitted by the many "centers of activity" responsible for solar stuff emanation at or around or near Earth: for example, centers of activity such as solar flares, CMEs, plages, facula, etc. Sunspots do emit, but so low it is almost negligible compared to what other "centers of activity" send forth. Sunspots are mostly a "sign" or a bearer, of what these more powerful emitting phenomena may bring: not exactly how much they do bring.

TSI then—the "solar constant" in old terminology—has a non-linked time relationship with sunspot number that is not singular but is multifaceted and complex, and still misunderstood and flat out unknown in many aspects. TSI and sunspot groups are independent and nonlinear in behavior in relation to each other and to the solar maximum-driving machine on the Sun, which is also known as the toroidal magnetic field (which will be analyzed in detail forthwith). As such, there is much that is unknown about what the "sign" of the one (the sunspots) means, versus what the others (CMEs, solar flares etc.) can in turn do. Sunspot number is but a proxy of this magnetic field and as such, points out the weaknesses scholars and scientists face in the use of this, or any other, proxy measure (such as the aforementioned C14 and BE10, etc.). This magnetic field is so far as is known as regards "mechanisms" a series of doughnut-shaped rings (or, toroids) that slide up and down and outward from the solar meridional center—that big line from the solar north "pole" to the solar south "pole" on the Sun. A sketch on how it might function—as error packed as this is—is the intended object of the next chapter.

---

[173]   For example, as the Air Force Research Laboratory (AFRL) SSN Workshop series is involved in (e.g., Cliver, E.W., "Why the sunspot number bears re-examination," Space Vehicles Directorate, Sacramento Peak Observatory, Sunspot, New Mexico)

Once again, the linearity of the Sun's behavior is often emphasized when it is in fact nonlinear in many ways. Our understanding of the Sun is fractured and in disarray. Perhaps unifying the various aspects of the Sun into a whole is impossible for our limited human understanding. All this has implications regarding the understanding of the effects its activity has, temperature, climate-wise, and especially electromagnetically in general, on Earth—whether in the long or the short term.

Now that we see how the Sun and Earth may interact in this constant dance of energy and matter, as error prone and half-known as it is, a more pedagogical look is taken at the Sun's energy emitters to get a better sense of what they are. That is, the Sun's so-called "centers of activity."

Isaac Newton

5. "Centers of activity" on the Sun: a linear view of the nonlinear as an introduction to helioseismology

I n attempting to define in the broadest sense of the term "activity centers" on the Sun, it is necessary to visualize the Sun as a gas that forms to our eye a sort of ball: but not a solid ball. The geometrical shape to our eye implies a solid yet it is not a solid. The sun is gas. We then picture this particular gas ball in the guise of a solid. All this pseudo-solid's breadth and depth we then see has specific areas of particle and wave emission. That is, certain select areas of the Sun have zones more prone to emitta and activity than others. All this belies a regularity of energy production that should not be there, would the sun be entirely chaotic.

It is somewhat illogical to assume that anything like a "center" is located on an irregularly-shaped pseudo-sphere of seething gas almost a million miles wide in (an imaginary, by the way) diameter. This however has to be comprehended and taken into account. Then we think what activities are going on in these movable and quasi-movable-loci. It is convenient to just label it activity at first, since not everything about the "activity" thus noted, and widely spread and nonlinear, is known. Words like plasma and particles and electro-magnetized fluids all come to mind and these are real in their labels but nothing widely and well known about any of it is commonly known. Thus do we hazard the pedagogical approach in linear tulip-labelling the Sun, merely to spread wide the potential avenues of understanding, if possible. Also rather lamely do we try and discuss motion in this chapter in connection with solar activity centers to spread a barely-visible conception of motion as regards the dispersion of the so-labeled "activity," and which will be looked at more deeply in the coming chapters. The distances covered here by the actual entity are immense. The "laws of sunspot" motion, as they come to have been called, must be looked at rigidly just to get a glimpse of semi-control in knowing how a star operates. For some of these laws describe essences of predictability that must be thought about very much more.

Ultimately, we risk doing to the Sun exactly what Aristotle once did to the layers of the atmosphere of Earth, somewhat like what was done in the last chapter. We cling to this linear shadow-knowledge in the hope of one day being able to do away with labelling (the ideal case) but which of course never will occur.

Perhaps the most familiar aspect of the Sun in terms of activity centers are one of the weakest activity centers of all: sunspots. Firstly, we note what sunspots and sunspot groups are. Then we note their relationship to motive fluid dynamics (magnetohydrodynamics), thence to electromagnetic behavior, as they show themselves, to observation, in Nature. Because sunspots' temperature is much lower than that of the surrounding solar "surface" (there is no surface, the Sun consists of gases and gases have no surfaces) sunspots and sunspot groups do not themselves emit much radiation. But sunspot motions are unique, somewhat predictable; are tied to dynamic inner and outer solar motion, and are visible surface "bearers"—or the magnetic "sign"—of solar "activity centers."

The Sun's "centers of activity" are, then, the "real" or greater emitters of plasma and particles. There are Coronal Mass Ejections (CMEs), solar flares, faculae, plages etc., on the Sun which are the strongest emitters of plasma and energetic particles from some activity centers. One main emitting "center of activity" is the CME. The enormous gas clouds that are released by a CME (more than a thousand billion tons of matter per CME) are called plasma clouds. They are called clouds since they carry electrically-charged particles in magnetic fields out into space. Another phenomenon, the explosive solar flare, releases the much-to-be-discussed Solar Energetic (or energized) Particles (SEPs). CME activity helps to spread or push SEP energetic particles to include Earth. As wavelength emissions such as UV and IR might influence the Earth's upper atmospheric chemical reactions occurring due to the absorption of light by atoms or molecules,[174] UV and IR emission bears implications for the composition of Earth's upper atmosphere like the ozone layer for instance. The electromagnetic fields carried along by CMEs along with say energized SEPs from flares of X-class can, if strong enough, influence electric fields on Earth. These carry portents for Earth surface-power line and oil line security lower on down in the troposphere, and all satellite function above the troposphere. That is, these particles, hurled at us so, and borne by CME magnetic wreaths, can effect advanced, widespread, electrically-dependent civilization nonlinearly and so, hazardously. How much harm, on the other hand, these can inflict is also nonlinear, sometimes localized, and depends on intensity of impact.

Secondly there is the rotational aspect to the Sun, talked about for now distinctly apart from fluid or current-bearing considerations as such, and purely as motion. One motion vital to widening this understanding is the solar motion in the lowest part of the Sun's convection layer. This relatively thin part not so thick but pretty deep down in the Sun is called the tachocline. The tachocline is so flexible it defies flexibility for all its thinness. The tachocline is the area where magnetic fields originate due to the following three variables of motion.

The tachocline:

1.   Contains the lower part of the outer solar convection zone (the convection zone extending from 700,000 km [c. 450,000 miles] from the solar center up to the photosphere)
2.   Is the lower part of the solar layers ("peels of the onion") where solar rotation increases in angular velocity (by contrast, the core rotates as a rigid body)
3.   Is where the solar rotation rate varies with solar latitude.

---

[174]   Or, photochemistry

Taken together, these three aspects to the tachocline conspire to cause gigantic whirls of electrically-charged particles which in turn produces the magnetic fields that make the Sun operate as an energetic entity.[175] For this, an understanding of magnetic dipoles is necessary (for a preview of what dipoles are in magnetic fields, see Chapter 6).

The CME/solar flare activities are linked. What usually occurs in Nature is often a solar flare eruption (of whatever class, from the weakest to the strongest—like X class). These are followed by CMEs and so work in tandem with hurling forth energetic particles. The size of a CME can be the diameter of 100 suns: the phenomenon is larger than the Sun by far. What are termed Halo CMEs are the ones directly effecting Earth, it is believed. The solar flare's or flares' heating is precipitated by the CME magnetic wreaths moving at incredible speed. Temperatures jump from 80,000 Kelvin (K) to 3-plus million K. We mention at this point the wavelength-dependent energies (Γ, UV etc.). For as the rush of magnetized gas and particles penetrates the hot solar corona in an adiabiatic plume, the CME speeds these outward. In an Archimedean spiral (soon to be discussed) this corkscrew punch moves with the solar wind at around a million miles an hour, ionizing plasma of SEP protons. "Extreme" UV (EUV) consists of 100 different emissions of UV radiation alone [176] and if part of a halo CME, will in effect create the ionosphere at the upper parts of the Earth's mesosphere. As was shown in the previous chapter, this modulation of UV in the stratosphere as it sinks down is what is thought at this point to be able to modulate—or be a very strong modulator—of Earth climate and weather. [177]

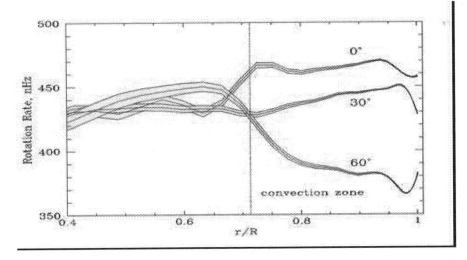

The tachocline (at the base of the convection layer) generates magnetic fields from electrical sources.

---

[175] There could be less dynamic forms of operation in the Sun.

[176] Tom Woods, SDO EVE project summary in 2010 (leader of EVE subsystem of the SDO)

[177] NASA's Living With A Star project outlines this and similar behavior thought to be exhibited by the Sun, using animation in some cases.

Deeper in the Sun at the core there is another motion, inertially driven, or, something comprising the solar inertial interior motion in the solar core (see Figure 3). In physics such a motion could be described as a thing going in a straight line until an external force makes it change direction: kind of an in-place shifting motion. This inertial motion portends much for what we know of as the cycles of the Sun in long and short time periods.

Thirdly, the three "peels of the onion," the photosphere, chromosphere, and corona, all emit from the "centers of activity" just mentioned at different wavelengths (see Figures 3 and 4). How and what these emit will be described here as clearly and distinctly as is possible. This description allows one to ponder any relation these "peels" may have to the solar rotational aspects previously mentioned. Keeping this in mind, one other aspect of activity centers that influences Earth magnetism is the Energetic Emissions Delay (or, the EED) factor and which plays a role in geomagnetic storms. What is glanced at here in this chapter is this case of how energy released from the Sun does not always show up as an effect for quite long periods of time on Earth. All of this should hopefully help us consider what is meant by "TSI" and "SSI" (Total Solar Irradiance and Spectral Solar Irradiance, respectively) more thoroughly. TSI is discussed in this chapter concerning some peculiar aspects of it, some wavelength emitta, and the chromosphere and photosphere. But TSI and SSI, along with SEPs, will be discussed more thoroughly in the following chapters regarding the solar dynamo.

◊

Sunspots mark—even define—what are called the centers of activity. These activity centers (like faculae, plages, solar flares, CMEs etc.) are the biggest "emitters." But sunspots themselves rarely are responsible for emitting any electromagnetic activity at these "centers of activity" themselves. As in telecommunications science, it is perhaps best to describe sunspots as "bearers" (the visible and logical sign of the emitting cables) of energy or the propagation of it: not as "emitters" (the physical emission). But again, the discussion of what emission is, here, is a relative and not very well known thing. Sunspots do emit radiation but in view of their relatively low temperature (~ 4000 Kelvin [K]) that emission is much less than that of the photosphere and hence, the spots seem dark. Their magnetic field strength can be as high as ~ 4000 Gauss, seldom are they higher. (This knowledge is surprisingly old [178]) Sunspots move vertically at 4000-6000 Gauss (see Figure 1). When they rise upward from the tachocline they drag weaker fields along (some 200 Gauss) and these appear as the solar activity regions. These areas are much larger than the sunspots. The areas' average sizes are 40,000 km (c. 25,000 miles) in solar longitude and 20,000 (c. 12,500 miles) in solar latitude. They show bright patches (faculae, plages) where magnetic energy is dissipated. These "patches" have temperatures in the order of 10,000 K.

---

[178]　American scientist (and Smithsonian Institution head) Joseph Henry noted that less radiation emanated from sunspots as early as 1845.

Sunspots you could say "live" from a few hours (the very smallest) up to days or even weeks and sometimes as long as a few months (the very biggest). The big ones move slower, live longer (sometimes up to several months and which can be the signs of weakened solar activity overall). Sunspot diameters range from between about 12,000-24,000 miles (or about two Earth diameters upward—or at mid-width, the Sun's tachocline). They have complicated velocity fields under and around them. Less energy rising up gives some a darker area (umbra) than higher energy ones (penumbra). This is logical, since most energetic "centers of activity" are mostly around the spots and not in, the spots. The magnetic fields (or, "sign") of moving groups are reversed, the lead spot with greater magnetic flux, and as such, resemble "centers of activity" in this sense. This magnetic field "sign" reversal is the "sign" of a new sunspot cycle according to the "second and third laws of sunspot motion."

Sunspots (called "macula" [179] in the ancient literature) to reiterate hardly emit radiation or particles: "centers of activity" around them do and are generally brighter than sunspots. Hydrogen A (H'α) reveals rising loops of connecting, oppositely-polarised magnetic fields. These fields in turn carry electrical current which propagates along the loops. The larger loops are "flux tubes" (up to 10-13 Angstroms [Å] [180] or 1-1.3 Nanometers [nm] in non-atomic measure) and have "flux threads." A crude metaphor would be a large copper cable consisting of twisted copper wires popping energy at the tops. But this is using conduction (a solid wire conveying energy) versus convection (fluid "wires" doing the same thing).

The tracks of the "centers of activity": sunspots as the "activity indicators" and their "laws" of motion

The 11-year sunspot cycle [181] was discovered by the German, Schwabe (in the 1840s, confirmed in 1851) as noted earlier. The length of this very visible and widely-used cycle varies between 8 to 13 years, the range thus being about 11 years.

The 22-year magnetic cycle (or, double, the range of the Schwabe Cycle, and sometimes called the Hale Cycle) was discovered by the American, Hale (in 1924) as "polarity laws" governing bipolar spot groups:

1. The first says that the distribution of polarities between leading and following spots is the same for all spot groups on one (northern or southern) solar hemisphere and that this arrangement is opposite to that on the other hemisphere or, one half the duration of the Schwabe Cycle.

---

[179]   Latin for "spot," much as facula is a Latin word for "torch."

[180]   The English spelling often omits the Swedish "Å" so it appears as "Angstroms."

[181]   Also called periodicities

Hale "polarity laws":

1. Distribution of polarities between leading and following spots is the same for all spot groups on one (northern or southern) hemisphere and that this arrangement is opposite to that on the other hemisphere = or, one half the duration of the 22 year Hale Cycle.

2. In the following 11 year Schwabe Cycle (say from 23 to 24) the opposite distribution of polarities applies.

3. *The solar polar field reverses between two successive Schwabe Cycles . . .*

"40 degrees or higher"
- **Unipolar Regions (UR)**
- **Ephemeral Active Regions (EAR)**
- **Polar Facular Regions (PFR)**

"Spörer's" (or "Carrington's" Laws of sunspot motion

"or lower than..."
- **Unipolar Regions (UR)**
- **Ephemeral Active Regions (EAR)**
- **Polar Facular Regions (PFR)**

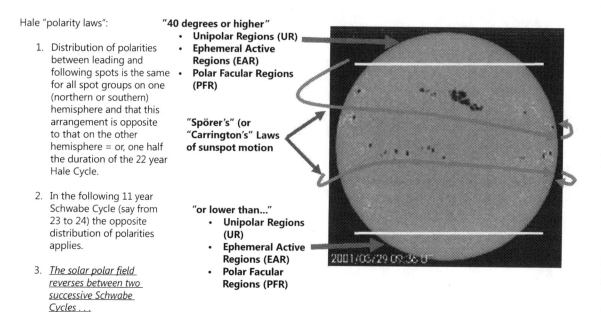

Figure 1. Laws of sunspot motion and the peculiar, separate nature of sunspot "bearer" motion versus polar spot activity—which appears to be separate.

2. The second says that in the following cycle (say from Sunspot Cycle 23 to 24) the opposite distribution of polarities applies. (If the solar north pole was UP in one cycle, the solar south pole is DOWN in the next cycle.) So, when sunspots from opposite polarity appear (negative leading) whereas the last sunspot groups were positive, say, then a new sunspot cycle has arrived. This arrival does not promise any set amount of sunspots, by the way, either at solar minima or maximum. (To get a better idea of what the phenomena is, see the sidebar on Hertz and his discovery of the magnetic dipole in Chapter 6.)

3. The third law says that the solar polar field reverses between two successive Schwabe Cycles (or, 22 years).

Spörer's (or, alternately, Carrington's) "Law" goes like this: during the course of a Schwabe Cycle, "centers of activity" are steadily found closer to the equator, starting at 30 degrees N. solar latitude, then twist off the east limb near the solar equator.[182] Maunder attached Gustav

---

[182] These ultimately give rise to the "butterfly diagrams" and E.W. Maunder who first made these diagrams and which coincidentally, illustrates Hertz' magnetic dipoles. The technical discussion of how butterfly diagrams arise out of the sunspot indicators within context of the de Jager-Duhau synthesis of the solar dynamo theory is discussed in the next chapter.

Spörer's name to the "law" in his honor as Spörer was perhaps the first human to witness such a phenomenon and remark correctly upon it. [183]

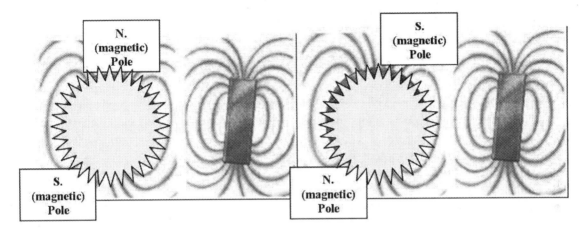

Solar polarity literally flips like a magnet.

Little did Spörer know how close he was to seeing the solar dipole in all this. But since fellow-German Heinrich Hertz had barely detected what a "dipole" was as an Earth phenomenon in Spörer's lifetime it is understandable.

George Ellery Hale, pioneer American solar astronomer (1868-1938). His invention of the spectroheliograph indirectly allowed seeing vortex clouds of possibly molecular hydrogen [184] near sunspots termed "flocculi" that is a potential "center of activity." (Mt. Wilson Observatory)

---

[183]   Yet English (technically amateur) R. Carrington's name is also attached to this "law."

[184]   Molecular hydrogen is the same variant found on Earth. (Most of the universe consists of atomic hydrogen.)

◊

We can think of the Sun quite rudely as if it were a mechanical thing, thus impinging on its majesty. Of course it is nothing of the kind and is indeed quite majestic. But mechanical terminology is useful for sketching the dynamics involved. The Sun rotates DIFFERENTIALLY (for purposes of this discussion, "in different ways, at different rates.") We have discussed the word before. Like a differential on a vehicle, this means that, whatever is in motion is basically going all over the place alternately constrained and assisted by pressure; sometimes all at once, yet at differing rates, and sometimes off and on—the off and on also at differing rates. Then, it restores itself to something they used to call the equilibrium state, but this is not strictly possible to do in Nature. It is something not witnessed by the average human eye and is difficult to comprehend withal.

So we consider the Sun in some rude mechanical way. Picture it if you will as a rounded Rubik's Cube where the inner part cannot be flexibly moved. But the outer can move or be moved very flexibly.

- The Sun's OUTER rotation is FLEXIBLE and CHANGEABLE as a "body."
- The Sun's INNER rotation is a RIGID BODY (yet, it moves "inertially").

The following description is the doughnut and ring metaphor in order to start understanding the phenomena. As mentioned in Chapter 4, animating this may help. But animation can distract as much as describe.

Figure 2 only shows the Sun's OUTER layer and how it moves. It is not showing a sphere: rather, it is describing motion, and the energy aspect is left aside momentarily.

The core is not shown or implied in this diagram (see Figure 3 for the core). Neither is the tachocline. The abstracted motion shown in Figure 2 is the result of what is in the tachocline. For this, take a look at Figure 3 or 4. In the OUTER convection layer upward to the surface (starting at c. 125,000 miles, a small solar distance) you find the tachocline "in" this 125,000 mile-wide part of the convection zone.

In the deceptively "skinny-kid" tachocline, approximately 20,000 mile thick, is found the "engine" powering outward magnetic loops to connect to the Sun's poles (the poloidal magnetic "fields" seen in Figure 2). The poloidal field off the solar axis and toroidal rings draped around them ("doughnuts") are made in the tachocline and are stored in the tachocline, like the expandable accordion.[185]

---

[185]   Invented, interestingly enough, by a close experimental associate of Michael Faraday. See Thomas Crump, *A Brief History of Science as seen through the development of scientific instruments* (Carroll and Graf: 2002)

In spite of its size, the tachocline is responsible for shooting out the fluid press (the motion outward of the solar stuff, out past even the solar corona) and at a very great distance into space. This fluid press creates among other things the heliosphere, itself. That is to include *all* the wavelength emissions and even the SEPs.

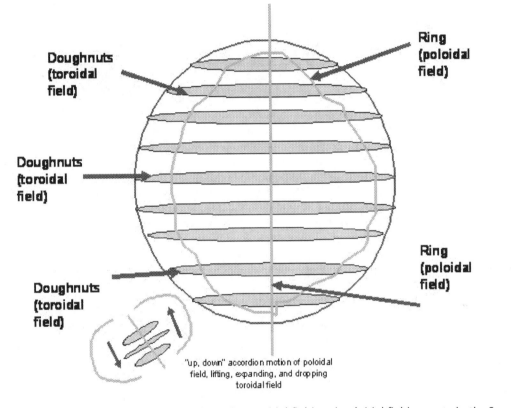

Doughnuts (toroidal field)

Ring (poloidal field)

Doughnuts (toroidal field)

Ring (poloidal field)

Doughnuts (toroidal field)

"up, down" accordion motion of poloidal field, lifting, expanding, and dropping toroidal field

Figure 2. The FLEXIBLE solar motion where the toroidal field and poloidal field operate in the Sun. The doughnuts (toroids) get pushed up and down and out, accordion style, along the poloidal ring's central axis.

As mentioned, the tachocline is only about 20,000 miles wide—less wide than many sunspot groups, and it is a bit larger than two Earth diameters (or, radii). So it is tiny in terms of width. But it circumnavigates the entire Sun's gaseous/particle/wavelength reach. Naturally, the Earth is affected by this spray of emission and particles though not of course linearly.

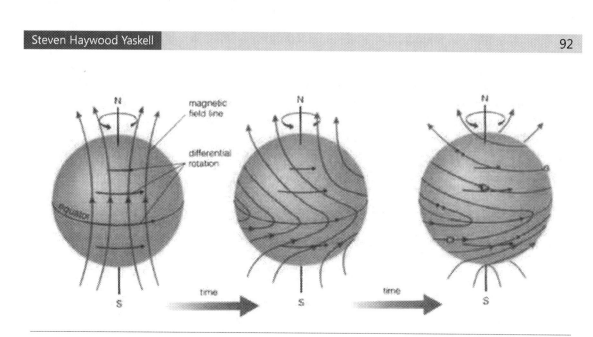

Differential motion and the Omega Effect: the Omega Effect changes poloidal fields (far left) into toroidal ones (far right). The Omega Effect is the magnetic field strengthening in a direction parallel to the solar equator. This leads to a field component that is mainly a toroidal one.

But back to energy for a moment. Magnetic fields generated in the tachocline are strengthened by differential rotation: the Omega Effect changes poloidal fields into toroidal ones.

What the sci-fi-like term Omega Effect means is that since fluid conductivity is not perfectly done in the Sun, the Sun's toroidal loop may diffuse through the fluid, disconnecting itself from the original poloidal field. This effect depends on the rotational velocity of the fluid. Fields increase in strength during subsequent rotations.

What with a "fluidly differential surface," and "rigid" (inertially rotating) interior, the problem then of course is detecting solar variability in clear detail at all. Let alone detecting it and correctly sorting it out.

◊◊

The Sun as a layered onion is now turned to, trying to recall all the while that the Sun is barely constrained by any layering, which always implies stasis. Here we risk making a world of spheres inside of spheres as did Aristotle.

Figure 3 shows the "onion peels" of the Sun in some detail that demands concentration: it turns your attention to it regarding wavelength emitters from the "centers of activity" in a specific way. These are the UV, Γ, and the anomalous X-ray radiation we have gone over many times in various contexts till now—to include Extreme UV (EUV) and the potential lesser quantity of IR.

Figure 4 shows these onion peels in dynamic action. Photosphere, chromosphere, and corona: all these separate peels of the solar onion facilitate different wavelength emissions from centers of activity they may cover. The photosphere "emits in the visible solar spectrum" (that is, the ball we see on a sunny day. Our eyes detect the yellowish mass as a spheroid). But Gamma (Γ) radiation may arise off the photosphere from solar flares. The chromosphere (found below the tachocline, and where flares and looping occurs) on the other hand, and further up from the Sun's surface, emits in Ultra Violet (UV) and in the anomalous Infra Red (IR) in perhaps some quantity. But Γ is also seen in the chromosphere.

A note on TSI variation in the photosphere and chromosphere—and Earth effects

The variation of TSI (sometimes just called the "total irradiance" or "solar irradiance") seated in the chromosphere's "centers of activity" has already been shown. [186]

But alas, then there's the details of the variations. Details of irradiance variations seem restricted to the UV band in the chromosphere. That is, irradiance variations in UV are only apparently seen in the chromosphere. Recent satellite observations [187] shows this irradiance emitted by the photosphere does not vary much across the solar cycle. It shows that the variable part of the irradiance comes mostly from the chromosphere (in the loops of the prominences, streamers, "plumes" etc.)—nothing much from within the depths of the Sun. Nor does it show the variable part of irradiance from the area right below the chromosphere—the photosphere. This chromospheric "stuff" does not directly hit Earth's troposphere—our common breathing space—as threshed out earlier, but interacts with Earth's atmosphere via the coupling of charged particles in an Earth stratospheric-ionospheric mix. Extreme UV (EUV) also remains longer in the Earth's upper atmosphere, as noted by the SDO.

---

[186] US Naval Observatory researcher Judith Lean

[187] In space, and so, accurately—not from the ground, which due to atmospheric disturbances—is not as accurate). The earth's atmosphere also absorbs all radiation with wavelengths below ~350 nm, and also a substantial part of Infrared Radiation (IR) which makes Earth-based recording sketchy.

X-ray in solar wind, off of corona (solar stuff 1)

UV, IR, some Gamma (Γ) (?) off chromosphere (solar stuff 2)

Γ off photosphere? (solar stuff 3)

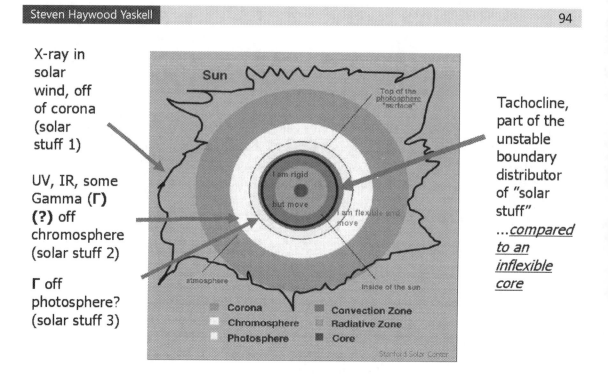

Tachocline, part of the unstable boundary distributor of "solar stuff" ...*compared to an inflexible core*

Figure 3. X-rays, UV, infra red, and gamma rays are all part of "TSI" and, from flares, come "SEPs"—the unstable, nonlinear tachocline being the flexible distributor of it out into the heliosphere with the aid of the corona and hence, at Earth. The tachocline, thin as it is, spans the Sun.

Then there is the solar wind off the solar corona [188] that is even hotter than the Sun's "surface" and is at a good distance from the corona. The electromagnetic spectrum's X-ray portion is emitted only by the corona, since the high temperatures found there (higher than the chromospheric "surface") do this.[189] Solar coronal regions at high solar latitudes, where the magnetic fields are open, allow plasma an outward flow (see the lower panel in Figure 5). Then the Sun's magnetic fields close only at a great distance from the Sun. In the so-called "basal coronal heating problem" [190] where the matter involves heat deposition at about 600,000 miles [191] from the Sun in the heliosphere (extended coronal heating out in the solar wind) one wonders why coronal holes and active and even quiet regions, as well as isolated loops, appear as varied as they do on the Sun. One reason could be "different combinations of mechanisms;

---

[188]   Delineated by E. N. Parker

[189]   In the solar corona the absorption coefficient is higher at extreme wavelengths.

[190]   Cranmer, S.R., "Coronal heating versus solar wind acceleration," Proceedings of the SOHO 15 Workshop - Coronal Heating, St. Andrews, Scotland, 6-9 September 2004 (ESA SP-575, December 2004). Ions have much more heat carrying power in this region of solar influence.

[191] R_0, or, solar radii where 1 solar radius = 432,450 miles.

for example, magnetic reconnection, turbulence, wave dissipation, and plasma instabilities." [192] The super heating of ions in this region pressed forward into Earth's upper atmosphere and how UV is modulated there is of great interest involving how Earth's climate may be affected, among other interesting topics relative the heliosphere and other stars in general.

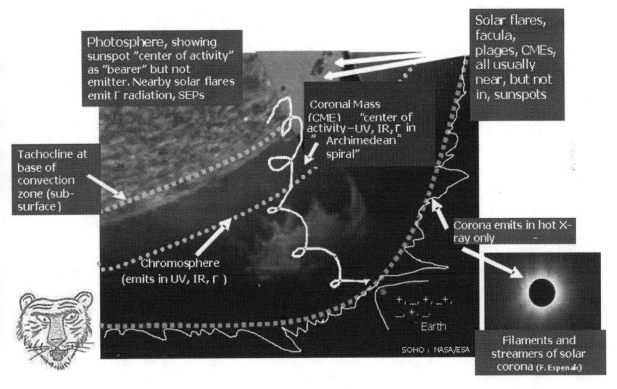

Figure 4. Wavelength emissions from the Sun are particular—that is, specific or even "locked"—to certain layers of the Sun's "onion peels" by the "centers of activity." These emissions come only from these centers of activity, are catapulted by the tachocline with SEPs, and are <u>not</u> part of the Sun's inner "inertial" rotation.

Centers of activity A: solar flares and "prominences"

Solar flares, the source of SEPs and solar-based cosmic rays [193] pop up when energy is suddenly released. [194] Solar flares emit extreme radiation lines such as UV, the anomalous X, and occasionally the Γ rays. A solar flare begins with "impulse phases" lasting from a few

---

[192]   Ibid, Cranmer p 154. The SDO project's subsystem HMI is examining the Sun magnetically from the core to the outer solar "atmosphere" to study exactly these phenomena.

[193]   Solar cosmic rays are generally defined as the particles with energies above some 10^8 to 10^9 eV.

[194]   This even happens from sunspots—only rarely

seconds to minutes. Heated plasma out of them can reach 50,000,000 degrees Kelvin (or close to 100,000,000 degrees Fahrenheit).[195] Cooling phases (that are gradual) which follow can last many minutes. These can be seen well in the prominences up and off the chromosphere (see Figure 5) and sometimes they are very dense. Reconnecting "flux tubes" explains flares' ignition, with only a few "flux threads" doing the igniting. At 40 degrees latitude up toward the north solar pole, polar facular regions start to dominate, as well as ephemeral active regions and unipolar active regions (see Figure 1). The "filaments" of flares have to reach these areas before they show up visibly on the surface. The "open solar flux" is a variable that is presently getting much attention. It is defined as the flux of particles leaving the Sun from a sphere at about 2 solar radii (c. 864,450 miles) from the Sun's center.[196] Open solar flux may eventually turn out to be an important part of Sun-earth climate considerations.

Center of activity B: Coronal Mass Ejections, or, CMEs

Altogether, Coronal Mass Ejections (CMEs) are the hydromagnetic releasing agents of massive amounts of magnetised plasma, or, "stuff" as it is still called [197] (10^{16} Gauss). They do accelerate particles at relativistic speeds into the Earth's atmosphere (see Figure 5, bottom image) to include the much-discussed SEPs from solar flares. During the CME's way outward into interplanetary space, plasma accelerated by the bow shock driven by the CME follow an Archimedean spiral magnetic field (see the "corkscrew" in Figures 3 and 4) that is in turn carried farther outward by the mysteriously hotter-than-solar-surface solar wind. The plasma thus ejected fills the heliosphere and disperses slowly. Co-rotational [198] interaction regions normally picks up the plasma at a confluence of weaker and stronger solar wind activations, and hence causes interplanetary shock waves, or bow shocks—something that will figure in prominently when solar orbital forcing is discussed in the next chapter. An analogy is to picture a four-armed water sprinkler that waves sprays of water around on a lawn, and the pressure you feel if you walk into these "arms" as they rotate.)

Then, the serious complication Nature poses here arises, which is to say that matter of the constancy of bow shock waving.

---

[195] However, studying the impulsive flare 1984/05/21, temperatures of 400 to 500 million degrees K in 13 pulses occurred *in a just one second*. These were the highest temperatures ever measured in a solar flare (de Jager, C., et al, *Solar Physics*, vol. 110, 317, 1987).

[196] At that distance—some take 1.8 or 1.9; more conventionally one assumes 2.5 solar radii—the field lines are more or less open.

[197] It is worthwhile to point out at this juncture that calling plasma "stuff" is perhaps "as clear" as Descartes' conception of "the aether" had been in his time.

[198] It is remarkable that Maunder noted a co-rotational solar aspect in the late 1800s (Ibid Soon-Yaskell: 2004).

There is in effect no time when the heliosphere (where the planets to include Earth find themselves inside of) does NOT withstand such shocks. That is, the heliosphere and its contents (to include Earth) are always withstanding shocks (see Figure 1 in Chapter 4). [199] CME plasma clouds collide with other gas clouds when released. During such collisions shock waves can and will develop. This again in turn can cause particle acceleration as in the previous discussion on SEPs, not to mention the CME's own charged particles. Such sped-up particles then obtain relativistic velocities. As such, with this magnetised rush of gas as it heads for our magnetosphere (see Figure 2 in Chapter 4) the CME's plasma clouds interact with it. They arrive from the Sun on the Earth's magnetosphere with velocities in the 300-1,200 miles (or 500-2,000 kilometers) per second range. What this then does is envelope the Earth's magnetosphere and compresses it many miles in the "tail" region (the Earth's magnetotail) in an anti-solar angle (as implied in Figure 2, Chapter 4). Due to such compression, the tail particles are accelerated and these move backward toward the Earth. Guided by the Earth's magnetic field lines, the particles then enter Earth's upper atmosphere through the magnetosheath. They enter the ionosphere as well, and through the polar regions Earth obtains its auroral lights (Aurora Borealis and Australis). Our planet also receives ionized particles that become scientists' proxy measurers of past solar behavior in this manner, such as the previously much-discussed isotopes of Beryllium, Carbon, and so on. These then end up in the Earth's lithosphere in one form or another (snow, trees, bones etc.).

---

[199] At Earth's distance the Southward Interplanetary Magnetic Field (nT) is measured at "6," and fluctuates by a factor of "10." Being a sort of descriptor of magnitude and direction, like seismic activity on Earth, nT is given simply giving a number (like the Richter Scale saying, "9" about the severity of the Japanese tsunami-earthquake catastrophe of 2011).

Phenomena near "centers of activity" (sunspots) as the Sun rotates, showing its "true spots"...

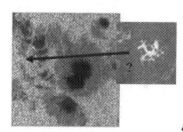

Plages ("beaches") (up, inset, uncredited); sunspots ("macula") showing cooler and hotter umbra and penumbra zones – they are "centers of activity" areas for radiation/magnetized particle emission

Facula ("torches") (Royal Swedish Solar Observatory, Grand Canary Islands)

...a magneto-hydrodynamic powerhouse that never stops

Solar flares (above, right) showing "coronal looping". "Prominences" emerge from such phenomena. A polar "coronal hole" (bottom) (SOHO: NASA/ESA; TRACE);

Figure 5. "Centers of activity" emitting on the Sun out at space (the heliosphere).

Relativistic particles of lesser energies than SEPs in turn become modulated by earlier-described CME bow shocks and can get fired at Earth in Archimedean spirals and are mainly Hydrogen—a substance that bodes much for the coming advanced research the Sun will soon undergo.[200] CMEs emit clouds of particles with an energy greater than 1 MeV (a multiple of an electron volt, or, a unit of 1,000,000 eV). They are coronal by nature and consist of Helium as well (and so, are mostly what the Sun is made of: much Hydrogen and some Helium [201]). Observed in Nature, and measured, intensities of up to 1,000 MeV have been observed.

One question regarding SEPs from solar flares and lesser-energetic particles out of CME bow shocks affecting Earth's electrical grids (etc.) directly comes into focus. Do the SEPs or lesser-energetic particles have bearing on power grids? The particles themselves and wavelength

---

[200] Being mainly Hydrogen, SEPs are also called solar proton events.

[201] About 71% Hydrogen and 27% Helium in mass

emitta are considerations here, but it is, rather, the magnetic fields (whether weak or strong) brought into Earth's magnetosphere during a geomagnetic storm that would have the greatest impact on power lines, satellites, oil lines, and so on and perhaps (as this is still nowhere clear) ultimately regional-hemispheric weather/climate. The magnetic fields are carried along by a CME. A geomagnetic storm is caused by the CME shock wave which interacts with the Earth's magnetic field. The increase in the solar wind pressure initially compresses the magnetosphere and causes an increased terrestrial magnetic field strength on Earth. In an east-west directed conductor, this voltage is small per unit of length. However, when these conductors are long, they eventually obtain significant values. Hence, for strong CMEs this may have dramatic effects, particularly in areas close to the magnetic North Pole (think of central and northern Canada)

The very curious case of delayed bursts of "solar stuff" . . . the Energetic Emissions Delay (or, the EED) effect

Very much more clear than what it does to Earth's climate is what geomagnetic storms prompted by CMEs' magnetic field manipulations can do to Earth power grids, geostationary orbital spacecraft, oil relays and so forth as it carries the power of strong solar flares.

The delayed, destructive action 48 hours after the initial blast in 1859 described in Chapter 2 noted by Carrington was the activity of delayed magnetic response. This delayed bombardment is not restricted to merely hours, but can extend to days, to months—and even to years. Once again we look at this possibility in relation to solar maxima. We think on the Gnevyshev Gap, a mysterious phenomenon whereby a solar minimum pops up between sunspot peaks.

The delay between the numbers of energetic emissions with respect to the time of sunspot maximum is called the "Energetic Emissions Delay" and like most things in electrical and aerospace engineering, gets awarded its very own acronym: EED. Carrington witnessed that September 1st (as a brewer, probably after checking his fermentation vats before lunch) what is now called a CME on his solar-photospheric monitoring 12 inch telescope. This was "a white flash" [202]—the electromagnetic effects of which were duly recorded at Kew Garden's Observatory 24 and then 48 hours later. Horrific effects were noted on the unsafe power grids at the time, blowing some up: burning others to the ground and spreading fires. This delayed effect could have been a type of energetic emissions delay, traipsing in hours after the "burst" all over the planet.

---

[202] Carrington was not absolutely the first, technically, recorded to have seen a "white flash." Pioneering British electrical investigator and (technically, amateur) solar astronomer Stephen Gray, alive at the Seventeenth Century's end, claimed to have seen something like this with his solar observing equipment in 1703. Papers on this exist in the Royal Society. A cloth dyer by trade, he was, at the end of a rude life, made an RAS member and given its first Copley Medal. Carrington's (and Gray's) flare hence are known as white flares. They were examples of a class of flares visible in integrated light without the need for an Hα (alpha) filter.

A fruitful research topic would be trying to define angles of width and breadth of "dead-on CMEs/flares" when such things occur to try and see at what angle on the night or day side of Earth could invite "the most destructive blast." But that would hinge largely on Earth receiving a direct hit from a specific solar latitude with a CME of such a strength as then witnessed. Given the non-linear nature of the Sun, this must by nature be rare and not always dead-on to cause the catastrophes often envisioned. The 1859 event, its apparition in Solar Cycle 10 (yet not particularly an overly strong maximum in the cycle) did indeed press back Earth's magnetosheath at a pin-point at some critical point on a cold, radiation-calm, near-3 A.M. Earth night side (c. 4 A.M.) punch and increased the atmospheric nitrates to a high level, as subsequently revealed by examination of proxy isotopes in ice cores laid down at that time.[203] Such nitrate derivatives, as shown in Chapter 4, can serve as the fuel for terrifically strong thunderstorms.

There is nothing exceptionally odd or strange about the events of 1859. Carrington observed the flare. It was certainly associated with strong high energy radiation (affecting stratospheric layers) and SEPs and solar cosmic ray particles that were part and parcel to the electromagnetic power that destroyed things in a delayed "punch." The SEP, cosmic ray, and CME-sent lesser energetic particles entered Earth's magnetosphere via the geomagnetic lines of force,[204] hence into the polar regions (see the diagram of Earth's magnetosphere, Figure 2 in Chapter 4). But their flux could not have been so strong as to markedly affect the Earth. The shock front of the associated CME that may have started at the same time as the flare reached the earth 48 hours later was the probable culprit. It compressed the magnetosphere and this compression helped cause the disturbances witnessed in for example telegraph line fires and burst primitive conductors, and so on way down in the lower troposphere.

There are other tales to tell in the case of what EEDs might be. It is a remarkable and unexplained fact that we often witness strong flares and associated CMEs a few years after sunspot maximum. The 1942 radio-burst which for the first time showed that the Sun can emit radio radiation was just such a case. Similarly, the solar cosmic ray burst of February 23, 1956 showed for the first time that the Sun can emit cosmic rays. The major flares of August 1972 and the so-called "Halloween flares" of November 2003 all came years after sunspot maximum and the reason why is still not understood. One guess is the tachocline, after having produced the usual solar activity, gathers a great deal of the remaining energy in a large explosion which later serves as the delayed effect.

---

[203]   Baker, D.N., and Green, J.L., "The Perfect Solar Superstorm," *Sky & Telescope*, Vol. 121, 2, February 2011, pp. 28-34. (See especially the bar graph of proxy data listed on p. 32 of this article.)

[204]   That comet tails always face away from the Sun was first thought to be due to a pressure from the gaseous emanations of the Sun. It was later found that this hydrodynamic "push" was due to a force quite strong in relation to wandering "dirty snowballs" of ionized gas—that is, comets.

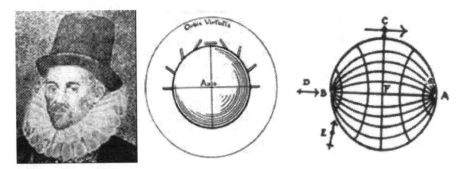

William Gilbert (1544-1603) English scientist and one of the first to experiment widely using magnets.
(Center) image of magnets behaving on a sphere. (Right) Noting the polar aspects of a magnetized sphere
(A and B). (De Magnete, 1600)

When considering magnetism tied to electricity for the simple semantic linkage of either concept into a complex and daunting whole—electromagnetism—the ideas and experiments of for instance the great English scientist William Gilbert and others become quaint. In the spectrum of light, electromagnetic, wavelength, gamma. X-ray, microwave, ultraviolet and infrared all take on a meaning of light guided or forced by magnetism are all similar if not particularly related. Solar particles in this stream, meandering into the lower Earth atmosphere, pose daunting questions. Stranger still is how Earth's very motion, its orbital characteristics, handle these entities in long and short terms.

For attempting to knit these fine points into a wider, stronger fabric as lucid or confusing as it might turn out be, we turn to the next chapter.

6. Total solar energy and particles, climate, Earth's orbital considerations, and the solar dynamo in the de Jager-Duhau synthesis

Like some phantasm hanging over us in our study of the Sun, the question of Total Solar Irradiance—what was once called the "solar constant"—has crept ponderously into our understanding. In this forthcoming excogitation of the solar dynamo as interpreted by de Jager and Duhau, the TSI/SSI conundrum stands out. As is beyond doubt, things which may conspire to affect our climate in a solar manner are due not to particles and waves alone, but in part also to the peculiarities of the Sun and Earth's motion and electromagnetic emitta.

TSI in relation to cosmic ray modulation, plasma, SEP particles etc. has implications for Earth climate and electromagnetic stability in any ultimate research sense if approached as a nonlinear entity. The difficulty of this is extreme although the mentioning of or insisting on it is easy enough. As mulled over in Chapter 5, the very details of total irradiance variations seem restricted to the UV band in the Sun's chromosphere; that is, TSI variations in UV are only *apparently* seen in the chromosphere, such as in the loops and in the prominences and streamers. We see nothing from "inside" or deeper down in the Sun, so far. Thus it is off the Sun's "surface" and outward in a sort of hydrodynamic press, out past the very hot solar corona (hotter than the Sun's interior and "surface"). Some of the particles hurled at Earth are strongly accelerated by magnetic field compression, as mentioned earlier. But only Iron (Fe) below the Curie temperature can be magnetised. This is not the case for any material in the Sun where the temperature is well above the Curie temperature.[205]

TSI is relevant to understanding the Sun's functioning in its "total energy" expenditure. As stressed in the last chapter however, areas of focus within this totality will yield positive results upon study. In considering TSI and any solar signature in Earth temperature as a kind of group, we think on TSI and its spectral side, Spectral Solar Irradiance (SSI). TSI/SSI in a newer understanding refers to attempting to comprehend the total electromagnetic energy of the Sun from the vantage point of clear space, from the rather low visible spectrum to very high energy (short wavelength components) such as UV and X rays. The first (low, in the visible wavelength) changes no more that 0.1% with the sunspot cycle. The second (high energy, say, Γ and X ray) might change 100% or even more from very low, to very high, solar activity. Thus it can strongly modify the Earth atmospheric composition and temperature from the stratosphere, upward. How it might modify Earth's atmosphere, *downward,* temperature and condition wise, is a major research issue and perhaps one of the most important ones. The Extreme Ultraviolet

---

[205] T_c. Magnetic effects are lost in elements heated above the Curie temperature, or, Curie point.

Variability Experiment (EVE) on the Solar Dynamic Observatory (SDO) mission at NASA has already shown, in the words of mission commander Dean Pesnell:

That most of the energy radiated (by a flaring region) does not consist of X-rays with wavelengths less than 7 nm (Nanometers) but at longer extreme ultraviolet (EUV) wavelengths around 27 nm. This has consequences far beyond our understanding (of solar flares). Earth's atmosphere absorbs EUV radiation at higher altitudes (say the stratosphere) [206] than it does X-rays, and EUV emissions last longer as well. Both of these observations indicate that scientists need solar spectral measurements in many wavelengths to predict the effects of space weather in our planet's atmosphere.[207]

Pesnell remarks in the same article that "extreme ultraviolet radiation from the Sun heats and ionizes the upper parts of Earth's atmosphere to such a degree that scientists call it 'the heartbeat of space weather.'" [208] It either creates or helps to create the ionosphere mentioned in earlier chapters. Note well Pesnell states that spectral measurements in many wavelengths must be done in order to note effects on our atmosphere. This in itself requires space-based tools free of atmospheric clutter to obtain a most difficult set of measurements. To say that it is very nonlinear should go without mention.

◊

TSI is, like centers of activity, a many-peeled onion. It must be examined in detail and then thought about in varying contexts.[209] For instance, it is now known that TSI has this spectral element leading to a totality of irradiance from a variable spectral source, relative to UV modulation in Earth's atmosphere (the aforementioned SSI). The next thing is to define and discuss TSI in relation to SSI, keeping in mind what has been ruminated on about solar behavior.

What about this totality in relation to us here on Earth? The Earth's main energy source is the Sun, the Sun giving off electromagnetic radiation with wavelengths covering the full electromagnetic spectrum and peaking in the visible part (at 400-750 nm). This alone should be all the proof one needs for claiming that the Sun steers Earth's climate. But such a claim is too broad. It is as scientifically specious as describing the Earth's atmospheric and the Sun's gaseous "layers" as the proof for either bodies' nonlinearity. What is Earth's climate, then,

---

[206]   Parentheses in this quote are by this author.

[207]   Pesnell, D.W., "Opening a new window on the sun," *Astronomy*, May 2011, Vol 30, # 5 p. 28. The SDO is an attempt to garner these spectral measurements in the many wavelengths, 24/7, for future analysis.

[208]   Ibid, p. 26

[209]   This interlude is gratis the paper, Beer, J. and van Geel, B., "Holocene climate change and the evidence for solar and other forcings." In: R.W. Battarbee and H.A. Binney (eds), *Natural Climate Variability and Global Warming: a Holocene Perspective*. Wiley-Blackwell (2008) pp 138-162. Footnotes are included from this paper in this book where appropriate to contextual discussion.

really? We take a stab at it in the dark. The Earth's "climate system"—a neat way of describing a vast, complicated mess from an aerospace engineer's view—one could define as *a system of pure energy transport.*

On its way through the atmosphere to the Earth's surface part of the Sun's radiation is reflected, scattered, or absorbed (refer to cartoon of the atmosphere in Chapter 4). The albedo effect is one feedback, as radiation (like short wave) is sometimes shot back out into space from glacial or rainforest "mirroring" etc., and the absorbed energy (about 70 percent) in any case is ultimately radiated back out into space in longer wavelengths. The openness of this our Earth atmospheric boundary is stressed, and it must be considered along with any arguments that this system can stay closed long enough to enact serious, catastropic climate change. Perhaps it can. Perhaps not.

As the incoming solar radiation covers only one half of the Earth (the day side) and peaks at low latitudes, permanent energy gradients on the Earth's surface are created. "Gradients" are the risings and fallings of temperature and pressure, variable as they are throughout the Earth's atmospheric "system," and down onto Earth's surface, or, its lithosphere and perhaps into water bodies. In this mix it moves particles in an ever-increasing amount downward. (Recall the discussion in Chapter 4 about "permanent" high and low pressure systems in an orographic context.) Due to the dynamic nature of Earth, its climate "system" naturally tries eliminating these gradients. What it does is shift them around in this process in a hyper-complicated shuffle the human mind and its attendant computational power cannot yet understand. Knowing let alone understanding the interconnectedness of such tradeoffs and transports in multi-million square miles of actual physical entities is beyond present grasp. The "system" transports energy received through the atmosphere's layers and even the ocean circulation. [210] The salinity of seawater plays a part in sometimes finding warm water under colder. Some "gradients" are sacks of water called thermoclines, wherein you will find warmer water illogically (according to laboratory interpretations of Nature) under cooler water flows. Ever swim in cool seawater only to feel warmer water under your feet as you moved? Due to static pressure laws, the warmer water should be over the colder, like warm air usually is (for example, in closed systems like homes in the winter). How these keep salinity for whatever purpose relative to global climate change was outlined earlier in the discussion on Henry's coefficient or law. But Nature does not always conform to static laws.

It is good to know that the radiative processes depend strongly on the atmosphere's composition. That is, all its gases, aerosols, the dustiness at the lowest levels, and the clouds there as well, as the Earth propels itself and is dynamically steered in space. All of this makes for more complexity, as these components are tied heavily to chemical, thermal, and dynamic changes taking place in the atmosphere on vastly differing time periods often confusingly

---

[210] A slowing of which can mean global cooling, as shown earlier in the 6,190 BC (c. 8200 years before present) in central Europe, in the middle east, and in China (see Chapter 4).

called time scales. And some "time scales" are very long, indeed, as we saw in discussing isotope reservoirs in earlier chapters. But in any case, the dynamic aspect of the Earth climate system means that these components will be moving around, or, constantly be getting moved around if not stored and then released. As we shall see the Earth is moved—is forced—about and around considerably in space.

Any change in these complex processes can in principle compel the Earth's climate to change.[211] Earth's climate change is not strictly altered by solar events, of course.[212] Here should come the text on orographical matters like volcanic activity affecting the atmosphere; for example, stratovolcanoes that pump temporary climate-changing sulphur aerosols and sun-blocking reflecting ash into the lower stratosphere. These aerosols block long wave radiation from the Sun, then fill the stratosphere and upper troposphere with sulphur aerosols that bring on the negative feedback of temporary global cooling. There is also the human-added gaseous waste that contributes to Greenhouse warming and which we do not consider natural. Both of these are more or less off-topic to this book. We reluctantly label all of these things with the (unfortunately) very ambiguous term *forcings*. We distinguish between external (orbital and solar) and internal (volcanic [213] and ocean circulation) forcings. In this book, we are limited to looking only at the orbital and solar variants of forcing.

Solar—and orbital—forcing of Earth

Why things happen to Earth differently each time as regards the Sun (and major planets) is due partly or wholly to differential motion. Planetary motions are, like all things we perceive, not perfectly precise.

Solar forcing. The amount of solar radiation arriving at any a given point on the top of Earth's atmosphere (indeed, upon ANY planet, such as Venus [214]) depends on:

---

[211]   One purely philosophical example of this is, should the fluttering of a butterfly's wings cause a perturbation in the atmosphere that grows so strong as to potentially effect extremely wide changes in atmospheric force?

[212]   Humans can and do contribute to climate change in local conditions in short time scales. How so hemispherically and permanently adding to runaway climate catastrophe is another question.

[213]   These are off topic to this book

[214]   Venus has been used as a model of what might happen to Earth regarding runaway CO2 effects. Had the Earth the same parameters for at least Venus' eccentricity and obliquity (leaving out precession) relative to *solar radiative absorption* and *energy transfer within* it, it could be a great concern. These are however not solar forces strictly speaking. Venus' retrograde motion also means that its north pole is almost always directly tilted at the Sun—whereas Earth is not so affected, not to mention Venus' decreased astronomical distance from the Sun's radiative emissions with its weaker magnetic field than Earth's. (Venus' atmosphere is mostly CO2, 96.5% by volume. Most of the remaining 3.5% is Nitrogen.).

- The solar luminosity (the total amount of radiation emitted by the Sun—determined by the Sun, only)

And:

- The relative position of the Sun and Earth in space (distance, direction of the Earth's [or even any planet's] rotational axis relative to the ecliptic)—determined by the distortion of the Earth's orbit by the gravitational forces placed upon Earth by the other planets (mainly Jupiter and Saturn). [215]

Orbital forcing. This affects three Earth "aspects" [216] with typical cycles [217]:

- The eccentricity (the deviation of the orbit from a circle, with periods of 400 and 100,000 years [218])
    - ✓ This one, eccentricity, means changes in the mean annual distance between the Sun and the Earth and so, to changes in the total incoming radiation from the Sun to the Earth.

- The obliquity (the tilt angle of the Earth's axis with a period of c. 40,000 years)
    - ✓ This one, obliquity, means that the relative *distribution* of the solar radiation on Earth is effected, impacting energy transport *within* the Earth climate "system".

- The precession [219] of the Earth's axis (c. 20,000 period).
    - ✓ This one, like obliquity, means that the relative distribution of the solar radiation on Earth also is effected, impacting energy transport *within* the Earth climate "system".

One "time scale" to focus on in these proceedings is the previously much-discussed Holocene (Chapter 3). The Holocene began about 11,700 years ago and continues, so far as we know, to

---

[215]   Laskar J., Robutel, P., Joutel F., Gastineau, M., Correia, A.C.M. & Levrard, B., "A long-term numerical solution for the insolation quantities of the Earth.", (2004) **428** (1), 261-285

[216]   "Aspects" is a term borrowed from astrology regarding angle, tilt, etc., as the heavenly body would look to the eye from a distance. This is one of the few remnants of astrology as transferred to working astronomy as initially promulgated by Johannes Kepler.

[217]   These are usually called periodicities in astronomy.

[218]   Those familiar with the pioneering work of Milankovitch and his cycles know about this figure range, and cycle, of a theory on deep ice ages (and is off topic in this book).

[219]   Amateur astronomers know this one from the current pole star, called (in Latin) Polaris in Ursa Minor. Each c. 20,000 years, a new pole star at the top of the Northern Hemisphere is obtained by precesssion, as the Earth changes its "aspect" naturally. The next new pole star should be Vega, in Lyra, way out in the future (c. 10,000 years from now).

the very second you read this.[220] Orbital forcing across the Holocene time scale is slow. But, the change in the total exposure Earth gets from the Sun's rays in solar forcing (in a less friendly and more technical way, called "insolation") over the past 11,700 is considerable. So it, too, cannot be neglected as an agent in "global climate change": that is, eccentricity, obliquity, and precession across all these years until this moment with solar forcing taken into account along with it. [221]

Solar activity and solar "forcing" of the Earth climate system: the limitations of the "solar constant"

Since the Earth's main energy source appears to be the Sun, solar variability is obviously a serious candidate for forcing climate change in some ways. We will not neglect mentioning geothermal, even human industrial, factors but relate that they would complicate this discussion beyond hope for the purposes of isolating the solar factor for serious discussion.[222]

Our Sun shows considerable cyclic changes in its magnetic activity, as shown by those bearers of electromagetic frenzy—sunspots and sunspot groups.

It was once demonstrated whether the solar radiation arriving at the top of Earth's atmosphere at a distance of 1 A.U. from the Sun was "constant." Or at least they tried to get a grip on this paradigm, somehow, by this method. Hence a "solar constant," or, what we have been terming Total Solar Irradiance (TSI) was created to build a kind of science archetype to wrap and rally meaning round, howsoever thin the understanding it yielded. But alas: solar forcing met orbital forcing. Most early attempts failed due to the above-mentioned radiative atmospheric processes on Earth, relative to dynamic motion (namely attendant eccentricity as to total amount received, and obliquity and precession, relative to the energy transports around the globe). Academic contra scientific understanding came to loggerheads, since this latter set of facts utterly complicated the TSI parameter. The major difficulty with the "solar constant" measurement was the absorption through the Earth's atmosphere. That problem can be overcome by measuring at various times of the day hence at various solar zenith angles. Since the intensity received varies with the cosine of the zenith angle, this allows for extrapolating to "zero atmospheric thickness." But it appeared that this extrapolation yielded errors far too

---

[220] This is the most recent epoch of the Quarternary Era, which also contains an earlier, cooler DEEP ice age period; the Pleistocene, where a lot of North American big game became extinct (Saber-toothed cats; beavers as big as bridges, and so on).

[221] Jansen E., Andersson C., Moros M., Nisancioglu K.H., Nyland B.F. & Telford R.J., 'The early to mid-Holocene thermal optimum in the North Atlantic." *Global Warming and Natural Climate Variability: a Holocene Perspective*, eds. R.W. Battarbee & H.A. Binney, (Blackwell, Oxford) pp 123-137

[222] That geomagnetic effects in the Sun-earth climate connection *alone* must be subtracted from understanding how much the Sun's irradiance ("the TSI") can directly effect Earth climate in short and long term time periods is difficult enough.

great for being reliable in any strict scientific sense.[223] To within the errors of these measures the solar radiation was "constant" ; hence the name "solar constant" now called TSI. Then the Space Age began, pitifully young still at this point. Along came extra-atmospheric satellites to give us all a clearer, and perhaps more frightening, picture. They were then able to mount radiometers on satellites outside the atmosphere. Only then was it discovered that the solar "constant" fluctuates—if in phase—[224] with the Sun's magnetic activity. [225] One had to wait for such space measurements in order to get a clearer understanding of what was going on with TSI and its spectral side. The database here is pitifully small, and missions like the space-based SDO are attempting to rapidly narrow the gaps in undestanding.

Let us now see if we can construct thought patterns detecting if the Sun is somehow, in some of its more important activity regarding us, more regular, then. We examine the centers of activity in this context. Measured TSI is broken down thusly into a:

- Background component
- Darkening component (controlled by the sunspots—that "sign" or "bearer" of the "centers of activity" [plasma/particles and wavelength emission] and seldom a strong emitter)
- Brightening component related to that "center of activity" still called faculae, or, "little fires" which overcompensate for sunspot's negative effects, leading to a positive correlation between TSI and solar activity. [226]

The satellite instrument data from c. the past three decades shows that the change of the TSI over a typical Schwabe Cycle is about 0.1 percent, corresponding to 0.25 W m−2, or, the global mean value at the Earth's surface. This is an estimate quite small compared with the 3.7 W m−2 estimated for a doubling of Carbon Dioxide.[227] This number regards what ill portents CO_2 increase due to human and other CO_2 production bodes for any potential global climate catastrophe.

---

[223] Personal communication with Cornelis de Jager.

[224] This word, "phase," takes on vital meaning in the context of radiative transfers in the concluding chapters. "Phase" is called 'anti-phase' if two major magnetic field components oppose—which is what is theorized as currently occurring on the Sun. De Jager, C., Duhau, S, *The Solar Dynamo and Terrestrial Surface Temperatures*, (ISSI: Bern), 2011

[225] The Solar Dynamic Observatory (SDO) is an attempt to begin measuring the Sun's multifaced emissions in a 24/7 way to offset this from the solar core outward.

[226] Ibid Fröhlich and Lean (2004)

[227] This amount of the latter substance is purportedly contributing to Earth's climate destruction in the Greenhouse Warming context. But we have gone over the oddness of the Carbon atom already, in its compounds, and how they store themselves in Earth's lithosphere, transmute, work into the relation of Henry's Coefficient and so on (see Chapter 4).

But the doubling of Carbon Dioxide causative for climate change is tackled by the variability of solar radiation, which is strongly wavelength-dependent. We recall Dean Pesnell's remark on measuring spectral emissions in many wavelengths to even start to understand the Sun's effects on Earth climate. Solar radiation reaches, as mentioned before, values of more than 100 percent in the UV part of the spectrum. Such large changes in the Spectral Solar Irradiance (SSI) portion of the TSI strongly influences the photochemistry of Earth's upper atmosphere—that is, the CO2 up there, and, in particular, the ozone concentration.[228] At least model calculations show that through dynamic coupling, the spectral aspects (SSI) changes can cause shifts in the tropospheric circulation systems and, therefore, change the climate. [229] But this has not been seen yet in Nature. It is a model. We must also remember or at least keep in mind that the figures just mentioned are global "mean" averages. And global mean averages (like [Δ] Delta T) reflect but do not describe Nature or its often to our experience brutal, capricious "processes."

Changes in SSI and TSI: on century or thousand-year time scales? Some climate aspects

From a climate perspective, changes in forcings on decade levels and on even shorter time scales are less important, because many processes within the climate system occur on much longer time scales (for instance, there's the thermohaline circulation). More critical matters on climate change revolve around whether or not changes of TSI and SSI occur on century and thousand-year time scales, and if so, just how large these changes in forcings, etc., indeed, are. They can be looked at like this:

- How variable is the Sun's magnetic activity? (This question is still as wide open as the universe.)
- How is this solar magnetic activity related to TSI and SSI? (The natural measure, versus the laboratory, or model, measure.)

---

[228] Ozone buildup in the upper atmosphere is a sign of less solar activity or seasonal winter effects or both. Ozone depletion is a sign of summer, or can be the sign of more Chlorofluorohydrocarbons (CFCs), man-made or natural, the natural ones being made by a reduction in solar electromagnetic activity (bromides and fluorides etc.) in upper atmospheric coupling. Human variants of naturally-occurring CFCs (like Freon) are thought less apt to be broken down by the Sun's activity or in any case have been overproduced and so contribute to the ozone layers' destruction, the layer construed as being a natural filter of skin cancer-causing UV. The other ozone layer is seasonally shifted or solar modulated or both (and the ozone hole is probably likely steered, and of course, will collect human pollution if too thick). The breakdown of ozone layers since the 1970s is thought to be a purely-human phenomenon or a sign of increased solar activity overall, or both.

[229] Haigh J. & Blackburn M., "Solar influences on dynamical coupling between the stratosphere and troposphere," *Space Science Reviews* (2006) 125(1-4), 331-3444

These are deep if honest questions that may never be answered. The magnetic variability is *larger* on longer time-scales, so far as we know. From the solar physics perspective, however, it is not at all clear if this is also the case for the TSI and the SSI.

Lending breadth to the arguments above are the paleoclimate reconstructions providing a growing body of falsifiable evidence for larger changes in solar forcing. They have been carried out for 30 plus years.[230] Not to belabor the points gone over in earlier chapters, but the longest human archive (and thus real) of solar activity is of the sunspot record, which (only) goes back to 1610 and the first savants who ever aimed telescope-to-Sun to observe and record them.[231] It shows the Schwabe Cycles recorded since then, superimposed over a generally increasing trend from 1610-to right now. And these of course are interrupted by distinct periods of low solar activity (for example, the grand Maunder Minimum of c. 1620-1720, and the not-grand Dalton Minimum of c. 1795-1820). And this 300 year record must be extended of course for a more intriguing view into the unremitting, invisible past, using, unfortunately, indirect proxy data. To remind once again, such data can be derived from measurements of cosmogenic isotopes [232] such as BE10 and C14 in natural reservoirs found in ice cores and tree rings, ("archives") respectively.[233] Cosmogenic isotopes (or, radionuclides) are produced continuously in the atmosphere as a result of the interaction of galactic cosmic rays with Nitrogen and Oxygen.[234] (See the sleeping and active Sun cartoons in Chapter 4.) So, the cosmic ray intensity is modulated by:

---

[230] Bond, G., Kromer, B., Beer, J., *et al,* "Persistent solar influence on North Atlantic climate during the Holocene." *Science* (2001) 294(5549), 2130-2136.

Haltia-Hovi, E., Saarinen, T. & Kukkonen, M., "A 2000-year record of solar forcing on varved lake sediment in eastern Finland." *Quaternary Science Reviews* (2007) 26, 678-689.

Neff, U., Burns, S., Mangini, A., Mudelsee, M., Fleitmann, D. & Matter, A., "Strong coherence between solar variability and the monsoon in Oman between 9 and 6 kyr ago." *Nature* (2001) 411(6835), 290-293.

Wang, Y.J., Cheng, H., Edwards, R.L., *et al,* "The Holocene Asian monsoon: Links to solar changes and North Atlantic climate." *Science* (2005) 308(5723), 854-857.

[231] Ibid, Soon-Yaskell (2004) and the early chapters on sunspot observers and recorders like Boyle, Cassini, Huygens, Hevelius, Scheiner, Galileo, etc.

[232] Radionuclides

[233] Beer, J., Blinov, A., Bonani,G., Use of 10Be in polar ice to trace the 11- year cycle of solar activity." (1990) 164-166

Stuiver M., Braziunas, T.F., Becker, B. & Kromer, B, "Climatic, solar, oceanic and geomagnetic influences on Late-Glacial and Holocene atmospheric C14/12C change." *Quaternary Research* (1991) 35, 1-24

Muscheler, R., Beer, J., Wagner, G., Changes in the carbon cycle during the last deglaciation as indicated by the comparison of 10Be and C14 records." (2004) **219** (3-4), 325-340

Magny, M., Solar Influence on Holocene Climatic Changes Illustrated by Correlations between Past Lake-level Fluctuations and the Atmosphere C14 Record, *Quartermary Research* (1993) 40, 1-9

[234] Masarik, J. & Beer, J., "Simulation of particle fluxes and cosmogenic nuclide production in the Earth's atmosphere, (1999)

- Solar activity

And

- Earth's geomagnetic field.

Ice cores are, as shown in Chapter 3, very good archives to measure BE10 and to reconstruct its production rate in the past. On the other hand, C14 forms C14O2 and exchanges between atmosphere, biosphere, and ocean. As a consequence of the large size of these reservoirs and the long residence times, the amount of an observed C14 change in the atmosphere is considerably smaller than the corresponding change in its production rate, and so, the C14 change is delayed.

But although the physics of the production processes in the atmosphere are becoming better understood, the transport from the atmosphere into the "archives" or reservoirs (ice, wood) is complex and has not been fully explained. [235] Hence we underline and accept the earlier reference to proxy measure as unfortunately limited. Had we several hundred more years of advanced science behind us, this would be less of a problem.[236] Proxy signatures suffer from the slings and arrows of a system of pure energy transport known as our climate. This is exactly where those who make sharp points about the unreliability of proxy data to explain solar effects on the Earth climate makes good scientific sense. Comparisons between BE10 and C14 records show that during the Holocene, the production signal was strong. This means that both BE10 and C14 provide an independent record of the cosmic ray intensity of the past.

But in order to extract the solar component from this production signal, the geomagnetic effect has to be removed.[237] This can be achieved by taking into account the changes in the geomagnetic field intensity derived from archeomagnetic and paleomagnetic measurements. The result is a record of the solar modulation function Φ [238] ($\Phi = 0$ means a completely quiet Sun, a condition that has never occurred and is just an ideal mathematical point, assuming no activity). $\Phi = 1000$ MeV corresponds to an active Sun as typical for a solar cycle maximum during recent times.

The "Φ-record" then is characterized by a long-term trend superimposed over shorter-term solar magnetic fluctuations. The shorter-term solar magnetic fluctuations pose the largest

---

[235] Beer, J., Muscheler, R., Wagner, G., Cosmogenic nuclides during isotope stages 2 and 3." (2002) **21**, 1129-1139.

[236] A major moral aim of all scientists is the obligation to keep enquiry free enough so that science will continue for several hundred more years to get a chance to obtain such answers.

[237] How this is to be done is discussed in this book's concluding chapters.

[238] Vonmoos, M., Beer, J. & Muscheler, R., "Large variations in Holocene solar activity—constraints from 10Be in the GRIP ice core." (2006)

problem for undestanding Sun-earth climate relations in an immediate, up-front sense. The short-term fluctuations can be divided into two pauses, or quasi harmonics; [239] that is, cyclic and episodic (and which will be described and discussed in the final chapter of this book):

- Cyclic
  - ✓ The Sun's *cyclic* [240] features the c. 11 year Schwabe Cycle, and the:
    - o C. 80-88 (?) year (Gleissberg Cycle, to include "pauses" in the magnetic Hale Cycle—more on this later)
    - o 205 year (the DeVries, or, Suess, Cycle)
    - o 2,200 year Hallstatt Cycle or climate cyclicity) [241]

- Episodic
  - ✓ Most *episodic* features are seen as strong, negative spikes corresponding to so-called grand solar minima, periods when the Sun was very "quiet" or as in the words of Agnes Clerke (in effect) "magnetically calm" (like the Sun in the Maunder Minimum).[242]

Φ Value

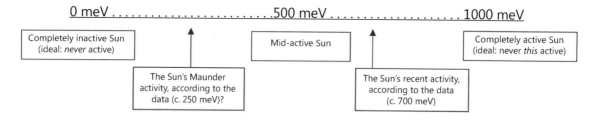

The ideal scale (Φ) of putative total solar magnetic flux forcings with the geomagnetic forcings removed (in MeV). Shown on this ideal scale is the Sun in the Maunder Minimum vs. the Sun in relatively recent times.

---

[239] Much more on these pauses in terms of harmonics will be discussed in relation to TSI etc., in this book's concluding chapters and indeed, is central to the premise of this book.

[240] Periodicities

[241] Stuiver M. & Braziunas T.F., "Sun, ocean, climate and atmospheric C14O2, an evaluation of causal and spectral relationships." (1993) **3**, 289-305

[242] It should be noted that this record does not cover the past 300 years. Compensation for this gap in reanalysed data will be covered in the concluding chapters, where de Jager and Duhau have included combed data to present their case for possibly predicting extended grand solar phases (maxima as well as minima).

The mean Φ value of the past five decades is about 700 MeV. This could be interpreted as follows: that relatively recently we were in a period of rather or even very high solar activity, ending as little far back in time as merely 2009. It also could be interpreted that in the far past—given mainly reconstructed sunspot records and proxy isotope data taken together—there were periods with considerably lower solar activity.[243]

Whether these large changes in activity are also reflected in the TSI and SSI is not known. Therefore, no speculations about the corresponding forcing in W m −2 are conjectured. (There is clear evidence for larger changes from paleoclimatic records, as discussed in Chapter 3.)

Scientists have used the ΔC14 ("delta") record as an indicator of past solar activity. However, Δ C14 reflects the deviation of the atmospheric C14 content, relative to a standard value, so it also contains a long-term geomagnetic and a system component which clouds up the solar "signal." Consequently the BE10 flux and the C14 production rate are better proxies, although they contain a geomagnetic component that needs to be removed, as well.

TSI measured as linear and TSI measured as variable and non-linear: two schools

The TSI factor: TSI taken as a fixed group and sunspot cycle motion, *don't* mix

For understanding the origin of the variations in the complete solar irradiance ("the TSI"—or the solar constant, even) you can also note that structural variations of the quiet photosphere over the Schwabe Cycle (differentially moving "surface") as seen by variations in both temperature and pressure depth-dependence (inner "stable" zone)—or both—have not been detected yet.[244] Keep this in mind apropos the earlier discussion revolving around Figures 1, 2, 3, and 4 in Chapter 5. Many voices and much research are still allowed to chop into this take on the solar dynamo theory and attempt to falsify it further.

---

[243] See also Solanki, S.K., Usoskin, I.G., Kromer, B., Schussler, M. & Beer J., "Unusual activity of the Sun during recent decades compared to the previous 11 000 years.", (2004) 1084-1087.

[244] The SDO mission is dedicating one of its subsystems in its three-fold plan to further understanding here.

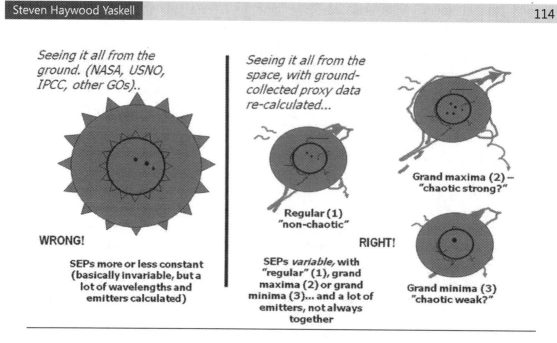

Tying SEPs (particles) to dynamo motion. Constancy is still a safety net to calculation of events that are separate and inconstant.

Why, for all the radiation from it we receive—it being the largest supplier of energy to the Earth and all—the Sun is assumed to have no large affect on Earth climate (especially in the short term) may be partly due to attempting to try and link the "R" (sunspot count) to energetic emissions using a certain approach. In a very difficult area, this approach is "too easy," as hard a thing as this is to say. And because it may be too easy (as difficult as it is) the fact that the Sun is the major bearer of our energy cannot be neglected. It might be hazarded that TSI / SSI is as differential as the Sun itself. Any attempts to measure it as a linear totality, as in some static number, against some parametric point, might work.[245] But it will not reflect Nature close enough to be a practical measure. The origin of irradiance variations could be ascertained from a study of the solar radiation spectrum and its variability. But, irradiance variations just cannot be detected from the "ground" (that is, from Earth) as accurately as is needed. It must be done from space, as we have discussed, due at least in part to the "solar constant's" fluctuation—in phase—with the magnetic activity of the Sun (in phase, in anti-phase, etc., in time). That the SDO is attempting to gather enough data from the Sun—from out in space—is itself a shot in the dark to attempt to see what is really going on in harsh Nature. [246]

---

[245]   Part of this is a ground-to-Sun observation problem. Fröhlich, C.: 2004, in: Pap, J. M., and Fox, P. (eds.), *Solar Variability and its Effects on Climate,* Geophysical Monograph Series 141, American Geophysical Union, p. 97

[246]   And interpreting this data, when it is available, might be aided by the use of radial basis functions of compact support.

The database for space-based irradiance measuring is understandably short and spotty, given that satellites for doing this reach back only a few decades and the collected sunspot record is also too short (if you can see c. 300 years as being too short. Chapter 1 was an attempt to ameliorate this). However, like Maunder with his missing climate data, we regret the shallow data bases that we must work with that are needed to increase our understanding of the Sun's actual operating mechanisms, by and of themselves, and in relation to Earth—even with the modern space-based measure missions now ongoing. Ultimately, more human time gathering and interpreting such data is needed. The Galileos and Newtons for this have not yet been born.

The TSI factor again: TSI *VARIABILITY AND* sunspot cycle motion *DO* mix

In this attempted understanding, the complications begin to arise considerably. First, there is no time lag between sunspot maximum and "irradiance" (TSI). This is evident from the physical connection between sunspots (maculae) and "torches" (faculae) (one the "bearer," the other the emitter). That there is no time lag between the maximum and sunspots and TSI supports the idea that sunspot numbers can be used as a proxy for irradiance variations. Furthermore, the solar radio emission at the 10.7 cm wavelength, which is often used as a tracer for solar variability, is primarily emitted by those layers of the "centers of activity" that are situated between a few thousands to some 6,000 miles above the Sun's surface (above the photosphere). And this radio emission varies in parallel with the UV radiance in the Sun's chromosphere.

Some metal gas emitters (Calcium, Magnesium) are well coordinated with TSI variations as well as with sunspot number, as measured by their leaps inside of magnetic structures. The conclusion, then, is that the fraction of TSI that does directly reach the Earth's troposphere is emitted by the solar photosphere (under the chromosphere: see Figure 3 in Chapter 5). It does not significantly vary, because the photosphere does not vary during the solar cycle.

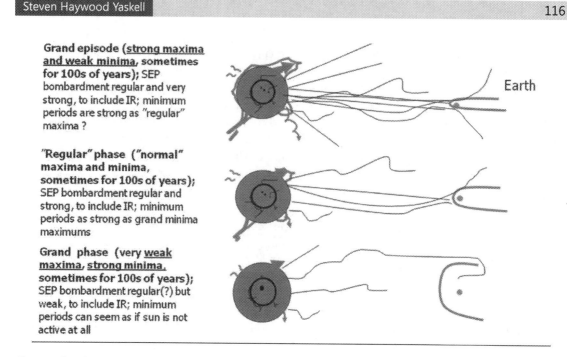

Grand episode (<u>strong maxima and weak minima,</u> sometimes for 100s of years); SEP bombardment regular and very strong, to include IR; minimum periods are strong as "regular" maxima ?

"Regular" phase ("normal" maxima and minima, sometimes for 100s of years); SEP bombardment regular and strong, to include IR; minimum periods as strong as grand minima maximums

Grand phase (very <u>weak</u> <u>maxima</u>, <u>strong minima,</u> sometimes for 100s of years); SEP bombardment regular(?) but weak, to include IR; minimum periods can seem as if sun is not active at all

Earth

Figure 12. Tying SEPs (particles) to dynamo motion. The Sun's SEPS behave in a non-linear fashion along the line of regular and grand phases.

The variable part of the solar magnetic radiation flux is emitted by the chromosphere's parts of the "centers of activity." And this flux only influences the higher, stratospheric terrestrial layers of Earth directly. This in turn influences the troposphere by some form of stratosphere—troposphere particle coupling [247] lower down. Coupling of particles in this context is that relation between electromagnetic variations and the gravitational force (the former related to charge, the latter, to mass).[248]

The solar dynamo in the context of Duhau—de Jager

Having noted the difficult aspects of TSI, SEPS and considered Earth's orbital restraints, we build a bridge between this regarding Earth's atmosphere and the energy transfers it handles from the Sun. This discussion should put the explanations for phenomena covered earlier into sharp focus, and was, indeed, promised in this book's introduction.

---

[247]   Coupling (as a constant) in particle physics is the measure of the strength of a type of interaction between particles, such as the strong interaction between mesons and nucleons, and the weak interaction between four fermions; analogous to the electric charge, which is the coupling constant between charged particles and electromagnetic radiation.

[248]   See Chapter 4 and the description of the ionosphere and the discussion in this chapter on Beer and van Geel's points for a review of the broader picture.

The solar dynamo theory is that the Sun's activity and its variations are assumed to be driven by a magnetohydrodynamic dynamo. This is of course quite a semantic mouthful. We will thus take it apart piece by piece. Magnetofluiddynamics or hydromagnetics, basically is:

the academic discipline which [is the study of the] dynamics of electrically conducting fluids. Examples of such fluids include plasmas, liquid metals, and salt water . . . The idea of MHD is that magnetic fields can induce currents in a moving conductive fluid, which create forces on the fluid, and also change the magnetic field itself. The set of equations which describe MHD are a combination of the Navier-Stokes equations of fluid dynamics and Maxwell's [249] equations of electromagnetism.

Note well the mention of force in these matters. Along with Babcock-Leighton models, we have MHD using "Navier-Stokes equations:"

The Navier—Stokes equations, named after Claude-Louis Navier and George Gabriel Stokes [mathematically] describe the motion of fluid substances. These equations arise from applying Newton's second law to fluid motion [250] together with the assumption that the fluid stress is the sum of a diffusing viscous term (proportional to the gradient of velocity), plus a pressure term. The equations are useful because they describe the physics of many things of academic and economic interest. They may be used to model the weather, ocean currents, water flow in a pipe, air flow around a wing, and motion of stars inside a galaxy. The Navier—Stokes equations in their full and simplified forms help with the design of aircraft and cars, the study of blood flow, the design of power stations, the analysis of pollution, and many other things. Coupled with Maxwell's equations they can be used to model and study magnetohydrodynamics.

Navier-Stokes equations, of course, can also be used to study the Sun.

We speak now about Navier-Stokes equations, Babcock-Leighton models, MHD, and the current-making involved in the solar dynamo—and then bring in Mr. (and Mrs.) Maunder once again. For what they discovered so long ago (in 1903) regarding "butterfly diagrams" as discrete but anomalous entities (that is, they did not know what they were) took until just recently to more-or-less lay out flat on the laboratory table for musing upon. But of course, this is based on observation and the math, naturally, yet by no means does it represent "the final word."

---

[249] James Clerk Maxwell's "most important achievement was classical electromagnetic theory, synthesizing all previously unrelated observations, experiments and equations of electricity, magnetism and even optics into a consistent theory. His set of equations—Maxwell's equations—demonstrated that electricity, magnetism and even light are all manifestations of the same phenomenon: the electromagnetic field." He was older than Kelvin.

[250] Newton's second law means momentum and mass in the combination of, say, like, when something is thrown at you and then it hurts after it bounces off. Force received, then, from this fluid motion's momentum is implied, nonlinearly, by such equations.

Figure 1 continues what we began in Figure 2 of Chapter 5—Figure 2 there describing only motion. If you recall Chapter 5, you will remember poloids (rings) and toroids (doughnuts) and tachoclines. That up-and-down accordion-like movement is shown to the left of Figure 1 as it operates in Nature.

The two loops on either side of the left in Figure 1 are the poloidal loops along the current. These flip up and out like electronic blasts from the monster's head in Dr. Frankenstein's laboratory. The flat lines in the middle of that SOHO shot are the toroids making the accordion motion, which blasts outward after the poloidal ring-shot like handles do on a jug. The output of this is an outward-flush of "solar stuff" could be only the wavelength emitter. (Maybe just the relativistic particles and EED effects). Maybe both if accelerated by bow shocks. And these go far out into space, into the heliosphere, and right past (and sometimes right into) Earth's atmospheric envelope, due to the thin but powerful solar tachocline.

We take the right side of Figure 1, stepwise, to describe what is going on according to Navier-Stokes, Babcock-Leighton, MHD, and the solar current:

1. Sunspot decay causes "loop" (poloid) field (smaller dotted lines) by Babcock-Leighton: there is a HIGH latitude magnetic flux from redistributed, low-latitude "centers of activity" flux—yet cannot re-create rotation in high latitude fields (not much visible activity at poles like UR, PFR, EAR. See Figure 1 in Chapter 5 as a reference as to what is going on here as far as observing sunspots is concerned).
2. The Sun's meridional circulation (solid lines) "carries" poloid field (large dashed lines) to the solar poles, then "sinks" it to the tachocline (at bottom of convection zone core: this is not shown). This sinking magnetic current takes the poloid field downward, latitude shear (rotational, at tachocline) and produces the . . .

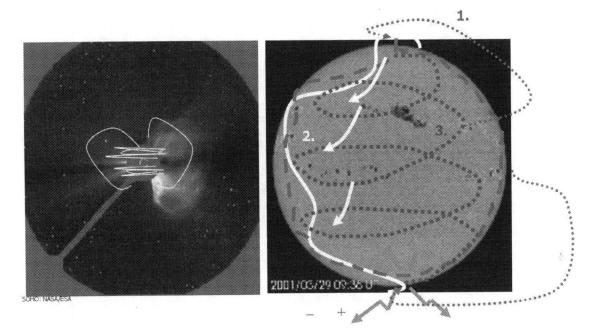

Figure 1. Loops and doughnuts to make current (in Nature, Left)—and a twisting (torsional) oscillation in tachocline as it proceeds, step-wise (Right) to create what is seen on the left. (Tachocline not shown: is in fact under surface by c. 125,000 miles)

3. . . . ."doughnut" toroid field (larger dotted line in "Archimedean spiral"). The toroid field rises to the surface due to magnetic buoyancy (the side-to-side shear in the tachocline ["the torsional oscillation"]) which breaks the magnetic field, thus producing current (lightning bolt symbol with arrows) [251] . . .

4. . . . which is described using Navier-Stokes equations. This allows the toroid (or, "toroidal") field to form at high solar latitudes of the tachocline without any surface eruptions. Alfvén [252] waves (low-frequency travelling oscillations of the ions and the magnetic field) then transmit the "magnetic stress" upward and outward from the Sun (to the "surface").

Figure 2 below is a continuation of what is ongoing in Figure 1 (right). Say we have stopped with step 4 (Figure 1) above, and now look at "step 5" (Figure 2) as a logical consequence of the induced conduction.

---

[251] Induction current, like a magnet and a coil. Faraday's equation has this produced from a side-to-side motion.

[252] After Hannes Alfvén, Swedish scientist whose name is attached to this magnetohydrodynamic phenomenon.

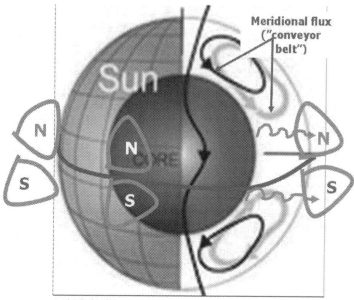

SOHO: NASA/ESA

Figure 2. Lorenz force and the "butterfly"("wings" marked North and South) stress.

5. The deep magnetic current, brought up to surface by Alfvén waves causes "Lorenz stresses"—which are N. and S. Hemispheric "butterfly wings"—or, sunspots bearing the "centers of activity" which in turn gives off the trace of their magnetized plasma action (the N and S-marked "wings" in Figure 2) on the equator-wards propagating activity of light and magnetism. (Maunder read out this data from the sunspot bearers of the activity in 1903 when he saw the wings, not connecting them to Hertz's dipole.)

6. A deep-penetrating meridian flow at the high latitude tachocline carries down into the stable, inertially-moved solar core a magnetic current that penetrates the Sun's depths. Alfvén waves then act (as in step 5). Torsional oscillation (the twisting) then happens before sunspots' eruption at mid latitudes. High latitude (EAR, PFR etc.) eruptions are powered *mainly* by core activity (and so, are less commonly seen.)

Given that the solar physics from the Lorenz stresses shown operating in Figure 2 (keeping in mind the activity shown in Figure 1 in this chapter, and those from Chapter 5) an explanation for the Maunder's "butterfly" diagrams of 1903 is seen, based on the dynamo theory in the de Jager-Duhau context. As a dynamic wave, it is measured routinely by NASA (see the top, right, of Figure 3). It is also a proxy of past solar behavior over many solar cycles and hence, many years and of course, is data for real-time-recording far into the future. When the Sun is weaker in activity, this dipole wave is somehow less distinct in the sunspot group counts that make up the wave manifestation (that is, they just are not shown in the usual c. 27 day counts that go into constructing the butterfly "wings"). Models have shown that when chaotically struck, the

Sun stops sending sunspots to the surface.[253] But that is a model and not Nature. The models in that paper [254] show "missing or badly damaged wings, relative to the solar hemisphere the spots have not appeared upon." Reconstructions of historical sunspot and sunspot group counts as done by Elisabeth Nesme-Ribes have shown the same thing (bottom right cartoon—not her own work—in Figure 3).

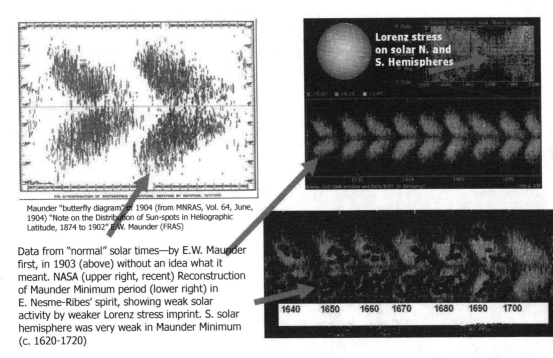

Maunder "butterfly diagram" of 1904 (from MNRAS, Vol. 64, June, 1904) "Note on the Distribution of Sun-spots in Heliographic Latitude, 1874 to 1902" E.W. Maunder (FRAS)

Data from "normal" solar times—by E.W. Maunder first, in 1903 (above) without an idea what it meant. NASA (upper right, recent) Reconstruction of Maunder Minimum period (lower right) in E. Nesme-Ribes' spirit, showing weak solar activity by weaker Lorenz stress imprint. S. solar hemisphere was very weak in Maunder Minimum (c. 1620-1720)

Figure 3. The actual, physical recordings of Alfvén waves, causes manifestation of "Lorenz stresses"—which are N. and S. Hemispheric "butterfly wings" of electromagnetic activity as revealed by sunspots and sunspot groups. Upper left: Maunder's recording these signals in 1903 from sunspot group manifestions over normal solar rotational periods (c. 27 days). Upper right: recent measurements of the same by NASA. Lower right: a cartoon of the late Elisabeth Nesme-Ribes' reconstruction [255] using proxy evidence of how these waves may have manifested themselves in the Maunder Minimum (using sunspot records from the time). That is: "less sunspots, less magnetic activity."

---

253   Tobias, S.M., *Astronomy and Astrophysics*, Vol. 332 (1997) pp 1007-1017

254   Ibid, Tobias, S.M.

255   The damning thing about her reconstruction thus was the spotty recordings of the sunspots themselves over those years, which heavily biases her work (for instance, "were there LESS sunspots during that time, or just LESS observers?"). However, keeping in mind Hoyt and Schatten's laborious historical reconstruction of who was actually recording these spots at the time, scientists like Hevelius, Boyle, Picard, Gassendi, Flamsteed and the like, the frustration is somewhat mitigated. See Hoyt, D.V. and Schatten, K.H., *Group Sunspot Numbers: A New Solar Activity Reconstruction, Solar Physics*, Vol. 181, No. 2, 491 (1997)

Any recent NASA recordings of these (say, in sunspot Cycle 24 and 25) should show, in time, altered butterfly wings and, if sunspots as bearers continue to be scarce, they should be showing badly altered "wings" in the real-time data.

What causes these wings of the Lorenz force to be more active? Some would say, "normal" solar activity.

But there are internal forces—hardly ever external forces—which conspire to force the Sun into going into "calmer periods" or, "anti-phase" magnetically speaking that have been quantified. To map the quantification we could use the mean Φ value, getting a lower Φ, where 0 MeV = no solar activity, and 1000 = extremely much solar activity. The number is not clear in these instances yet it is never absolutely 0 (an ideal number). But way below the recent 700 MeV is possible in any coming deep solar minimum.

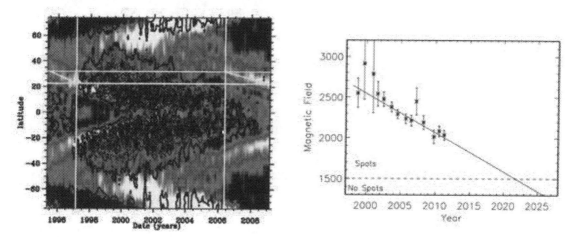

(Left) A helioseismic map of the solar interior (NASA). The east-west flow starts to form toward solar equator (20 to zero latititude and so, the visible sunspot formation in the coming sunspot cycle "normal" or "regular" sunspot maximum). The unipolar regions at the top of the Sun are where the "jet stream" preceding sunspot formation is located (c. 40-60 degrees N, see Figure 1 in Chapter 5.). The "rush to the equator" should be happening in the oscillations, but is not. (That is, it should have started in 2008). What is going on? Will this form and eventually reach the solar equator? If weak, would this force in the stream be recorded as broken or fuzzy in terms of sunspot amount as described above in the Nesme-Ribes' reconstruction? (Right) The magnetic field of the Sun has been decreasing since 2000. If below 1500 Gauss, "no spots" will appear in the listed future dates. (Matthew Penn and William Livingston)[256]

[256] See Pasachoff, J.M., and MacRobert, A., "Is the Sunspot Cycle About to Stop?" News Notes, *S&T* September, Vol 122 # 3. pp. 12-13. See also Salazar, J., Frank Hill: Future sunspot drop but no new ice age (EarthSky Interview, April 12, 2012) online. The original paper outlining the discovery of this polar oscillation was by

The Higgs Boson and plasma as a possible fourth state of matter

We talked much of the SEPs and the TSI/SSI, the particles viz À vis the wavelength and spectral aspects of the Sun. This is somewhat part of the account of the dichotomy of the particle-versus-wave phenomenon described earlier. What is worth thinking about here?

We recall the great Kelvin and try to capitalize upon his wonderful errors. The field strength of the Sun is in itself much slower than Kelvin had originally calculated. [257] Rolling quickly fast-forward to the mid to late 20th Century we arrive inevitably at the recondite and anomalous "Higgs field." Not a cricket or rugby field but a science one, the British physicist Peter Higgs [258] suggesting a wave-particle duality in the form of what is termed nowadays the Higgs Boson, which boldly in his theory suggests a field comparable to that of the electromagnetic ones we find in space. All particles that pass through it will acquire mass.

We directly quote the duc de Broglie:

> "If there is a particle-like aspect of light, there must be a wavelength aspect to matter"

> Duc de Broglie

What gives particles their mass? The wave-particle eternal transmutation, back and forth, needs the "Higgs Boson" to be the cause of this wave and particle protean relationship. This harkens, somewhat, back to J.J. Thomson's and Ernest Rutherford's work in the discovery of Alpha (A) and Beta (B) radiation. Rutherford found both according to their rates of decay, the A particle being a positive charge and the B particle being a negative one. Beta (B) particle loss (electron loss) increases atomic numbers in isotopic interaction, but does not affect the atomic weight. [259] The "weak theory" (force) here [260] is related to beta decay, solar energy creation, and superconductivity. The Higgs Boson would have elementary particles doing the

Hill, F., et al, "The Solar Acoustic Spectrum and Eigenmode Parameters" *Science*, Volume 272, Issue 5266, pp. 1292-1295 (May, 1996)

[257] Tragically the "father of" modern thermodynamics erred remarkably in his estimations of the Sun. "Mathematically, he wrongly assumed that the field strength of the magnetic waves decreased as an inverse of the distance to the cubic power. It is actually less rapid and perhaps (is) an inverse of distance squared." Ibid, Yaskell (2008)

[258] An update on the possibility of this particle's existence is from Dennis Overbye, "Particle Accelerators Full of Fury and Signifying Something," NYT, *Space and Cosmos*, 1 August, 2011

[259] In another take on the strange world of Carbon, with, it lets its radioactive isotope C14 transmute eventually (60 Kyr) into 14N, the *least* radioactive isotope of Nitrogen.

[260] Or the GWS Model of Glashow, Weinberg and Salam

weak interactions between it and the electromagnetic field (see the diagram of elementary particles below).

Is it perhaps in this strange world where we will see TSI and SEPs as "less separate" phenomena?

TSI and SEPS are separate phenomena so far as we know for now—if both nonlinear, and both a product of an unstable, open boundary system known as the Sun—which is also theory.

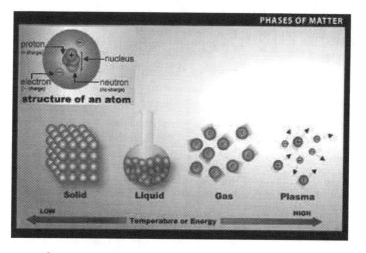

The familiar three states of matter alongside plasma, being a possible fourth state of matter due in part to the high temperatures involved that can only be measured in Kelvins. (Courtesy UC Regents)

The days of Fourier, [261] Navier, Kelvin—even Einstein—perhaps are over and we are approaching a new frontier of high energy physics that require new Kelvins, Rutherfords and Einsteins. Another kind of mathematics altogether perhaps must be used to better or more effectively show the solar dynamo functioning in harsh Nature possibly in this regard, along with newer physics, and using all the gathered data. Then we should be given over to interpreting any climate modulation of Earth by the Sun, and the effects such modulating phenomena have on very practical problems like power grid and fuel line protection, and satellite and space travel safety as well as, I personally hazard, agricultural-yield security. Aiding in increased harvests of solar power should definitely not be neglected as an electric generating source.

---

[261] Fourier also came up with what is termed the Greenhouse Effect, to which he applied convection currents in open systems. The "greenhouse effect" on Earth applies to closed systems—like actual greenhouses, for instance, where convection can be controlled, as well as humidity, etc. A "runaway greenhouse effect," the bane of some, on Earth, is likened to what occurred on Venus and is feared to happen on Earth should human CO_2 production not be reduced (all re-absorption reservoirs aside of course). As shown, however, different orbital tilt, eccentricity, Sun-locational and other factors conspire to show that Earth is not like Venus.

The new math. In the words of Duhau and de Jager:

Modes of oscillations of linear and stationary bound systems are harmonics functions; therefore any oscillation in these systems is (sic) well represented by the Fourier base function. The same is not true in non-linear systems with a non-stationary boundary for which the modes of oscillation change with time in frequency and in amplitude. For describing oscillations in such a kind of system, a base function of compact support is needed.

B. Table of Elementary Particles

Name	Symbol	Mass (MeV)	Charge	N_c	
Spin 1, gauge photons:					
Photons	γ	0	0	1	$U(1)$
Vector bosons	Z^0	91,188	0	1	
for the	W^+	80,280	+	1	$SU(2)$
weak force	W^-	80,280	−	1	
Gluon	A_s	0	0	8	$SU(3)$
Spin 0, Higgs:					
	H^0	>60,000	0	1	
Spin $\frac{1}{2}$, quarks:					
I up	u	5	$\frac{2}{3}$	3	
down	d	10	$-\frac{1}{3}$	3	
II charm	c	1600	$\frac{2}{3}$	3	
strange	s	180	$-\frac{1}{3}$	3	
III top	t	180,000	$\frac{2}{3}$	3	
bottom	b	4500	$-\frac{1}{3}$	3	
Spin $\frac{1}{2}$, leptons:					
I e-neutrino	ν_e	≈ 0	0	1	
electron	e	0.510999	—	1	
II μ-neutrino	ν_μ	≈ 0	0	1	
muon	μ	105.6584	—	1	
III τ-neutrino	ν_τ	≈ 0	0	1	
tau	τ	1771	—	1	
Spin 2, graviton:					
	g	0	0	1	

Base functions of compact support is perhaps best put by Martin Buhmann, a pioneer of a form of mathematics that is less than fifty years old: [262]

[262] Buhmann, M. D., *Radial Basis Functions: Theory and Implementations* (CUP: Cambridge Monographs on Applied and Computational Mathematics) 2003, p. 9

Radial basis functions are . . . useful especially when the number of data and evaluations of the interpolant is *massive* [263] so that any basis functions of global support incur prohibitive costs for evaluation . . . Moreover, radial basis functions with compact support are suitable for, and are now actually used in, solving linear partial differential equations by Galerkin methods. There, they provide a suitable replacement for the standard piecewise polynomial finite elements. It turns out that they can be just as good a means for approximation, while not requiring any triangulation mesh, so they allow meshless approximations which is easier when the amount of data has to be continuously enlarged or made smaller. That is often the case when partial differential equations are solved numerically. By contrast, finite elements can be difficult to compute in three or more dimensions for scattered data due to the necessity of triangulating the domain before using finite elements and due to the complicated spaces of piecewise polynomials in more than two dimensions.

[264] **Sidebar: a brief history of Maunder's butterfly diagrams and reverse polarity: the magnetic dipole field**

Electromagnetic butterfly waltz

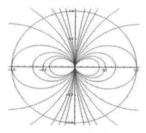

A magnetic dipole "field"

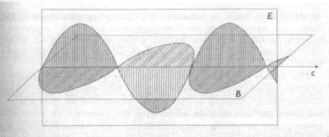

Figure 7.2   Electromagnetic field of light. A pencil of light moves in the direction c: the associated electric (magnetic) field vibrates in the plane marked E (B).

Heinrich Hertz

Discoverer of radio waves Hertz tested Maxwell's equations in the 1880s and devised an apparatus using a Faraday induction coil making a current along two spheres, one positive, one negative. In passing a gap, the energy of the electric (spark) current fades, with its energy being transferred to a magnetic field, moving at right angles. When the magnetic field collapses (as it will) a reverse electric current is produced: ie, what was a negative polarity becomes positive, and vice versa

---

263   Italics are by this author.

264   Figure 7.2 in the sidebar is adapted from Crump, T., *A brief history of science* (2001)

7. What a grand solar phase mechanism might reveal in the short term

In spite of the fact that the Sun is an unpredictable entity one likely—yet perhaps unreachable aim—is as follows. There could be some sort of repeatable predictive index into how grand minima and maxima form; perhaps even when grand minima as well as grand maxima could occur. That is, within the "normal" Schwabe and the less fathomable Hale Cycles, the "grand" variant (or variants) will sometimes stand out and be open for better prediction. This index might be arrived at due to solar behavior as recorded in the past as well as quantitative analysis re-made after observations from Nature using newer methodology and newer mathematics. The new mathematics are suitable to near-limitless computational iterations. The processes to achieve these will be the ones described earlier in Chapters 5 and 6.

We have seen how our variable star goes in and out of its "normal" minima and maxima phases and how some minima and maxima are "deeper," or last "longer," than others at times. It is important to note that these longer solar minima and maxima are tied to longer cycles. And so with this, the discussion begins. What we have already seen are exactly these long-term solar cycles as they change in time (as mentioned, the conditionally "longer term" Gleissberg Cycle, the quite long-term De Vries Cycle, or, Suess Cycle and so on.) Cyclical change in closed linear systems (like in laboratories) possibly occurs similarly in non-linear (non-isotropic) open boundary systems such as the Sun. Of course the scale is very different and so, difficult to comprehend. Yet, as we can map change in laboratories, so might we be able to map what changes occur in the Sun, as unpredictable as it is in its behavior. This, even if is much bigger than anything possibly reproducible in Earth-based labs.

The cycles are there, then. But then they vary and behave in as non-linear a manner as does the Sun itself, almost as if they were separate. Peeling back the onion layers of cycles we consider the concept of abrupt changes in time concerning all solar cycles, and especially the longer-term solar cycles. Let's think of an abrupt change that is easily pictured. In a stunt-filled movie you see a driver shift a gear in a car or a truck smoothly on a long, "regular" drive in the country, say, and the differential in the vehicle performs its unseen twists and turns in its mostly-forward motion. Then the action starts and the driver suddenly shifts the vehicle in reverse for a moment at a "regular" speed when the gears least expect it, then attempts to put it back into say second gear to third to obtain a "regular" speed again. The result here is a pretty chaotic occurrence of course. Abrupt changes of overlapping long term solar cycles (like the Gleissberg, say) cause what are called chaotic transitions. In the case of the car, the driver was the cause of the abrupt change in the gear cycle shift. In the case of the Sun, the abrupt-change actor of the long-term cycle shift could be the internal inertially-moved core. Higher up in the Sun, torsional oscillations stabilize the tachocline-convective layer motions. This in turn leads to a transferral of angular momentum in the latitudinal (the horizontal) direction from the core spin to the tachocline. This, then, could be the "punch" of chaotic abruptness sometimes witnessed in Nature, and in the data, or in both—on the Sun. Next, we think of the actual "transition point" itself. Think of the transition point alone as the total moment the driver of the vehicle caused the sudden gear shift just mentioned: either a regular shift to reverse or a reverse to a regular, shift. The transition point is the topic of this entire chapter and has as difficult a context here as the word "differential" does to the description of how the Sun operates in the solar dynamo theory.[265] We have a thin lead in this detective story of the Sun's move toward such transition points. These are in the repetitive and so fairly-predictable cycles themselves. We look at "all" [266] our solar cycles that have been mentioned in this book so far. We then list these cycles in a simple numerical order starting from 0 years to about 3,000 years, as we know them to go thus far in our limited understanding. These cycles are categorized for convenience into two rough groups in order to make basic distinctions:

"Shorter-term" sun cycles:

- C. 11-year Schwabe Cycle, the most "regular" of the known solar cycles (1/2 the Hale Cycle—the Schwabe's magnetic component)
- 22-year Hale Cycle (added with its visible Schwabe c. 11-year phase)
- C. 88 year Gleissberg Cycle (and its connection to the Hale and possibly to the De Vries Cycle) ignoring the upper and lower Gleissberg limits momentarily.

Longer-term "sun" cycles:

- C. 205 year DeVries (or, Suess) Cycle [267]
- C. 2,250-2,300 year (the Hallstatt Cycle) [268]

de Jager and Duhau took the c. 300 year solar cycle/ human-sunspot recording database (from c. Galileo on) [269] and the reconstructed sunspot recording data, along with the proxy (BE10,

---

[265] The appreciation of the word, "chaotic" is, in this context, a physical-scientist's interpretation of the word rather than a social scientist's. For example as found in Lorentz, E., *The Essence of Chaos*. University of Washington Press, Seattle (1993)

[266] Or, at least the most relevant for our discussion

[267] A C14-measured cycle not related to temperature that shows a sinusoidal curve, somewhat reliably describing long-term solar variability along that curve (minima and maxima) over long-term time

[268] A O18 and O16-measured cycle related to temperature, as opposed to the Suess. (We omit the recondite, much longer-term Milankovitch variation of c. 100,000 years from this discussion.)

[269] Literally, recompiled and cleaned data from random observations in this case, all the way back to the 1600s when observations of such activity were less organized, overall. Such measurements did not become more common until the 1870s and much more so thereafter. As this data and these proxies in some cases depend on atmospherically-disturbed, yet solar-produced isotope behavior, they do *not* tell an exact,

etc.) data. Doing this, they saw (as is illustrated in Figure 1) that a function of two solar physical relations emerged in generating this data. That is to say as functions of each other, they made transition points in time. But it is important to point out that this examination is related only to long term cycles; yet, the beginning of this actually starts in the Hale Cycle (a relatively short term cycle, tied as it is to Schwabe Cycles) and the Hale reaches into the Gleissberg (c. 88 year) Cycle and possibly beyond. This represents the "doorway," so to speak, for elucidating the aspects of longer term cycles involved in transition making. The function for "regular" solar MAXIMA are the TOROIDS ("doughnuts," plotted as "Rmax"[270]). The function for "regular" solar MINIMA are the POLOIDs (loops, plotted as "aamin"[271]). Both these functions are actual, measured solar magnetic-field components. So this implies instantly X and Y coordinate planes and other graphing methods upon which to plot the functions in past time (and into the future). The transition point is found from the behavior of a long-term component (see Figures 5 and 7). This long term component is defined as the sum of the linear trend and the wavelet component cycles in the Upper Gleissberg [272] (72-100 years) and in the De Vries Cycle (200 plus or minus years). Thus it seems as if the amplitude modulation of the Hale Cycle is further split into the Gleissberg and the De Vries Cycles.

---

to-the-year picture of what the Sun was doing, once, when. So it can justifiably be considered weak or tainted. Use of proxies in say meteorites containing isotopes, like those from Argon and Titanium, on the other hand, do not show this atmospheric "bias" and are considered more exact recorders of at least cosmic if not solar behavior.

[270] The data for Rmax at varying points in the diagrams to follow are from Nagovitsyn, Yu. A., "To the Description of Long-term Variation of Solar Magnetic Flux; the Sunspot Area." *Astron. Lett.*, (2005) 31, 557. The Rmax data from before and after 1705 are the Wolf and the Group Sunspot Number, in Hoyt, D. V., Schatten, K. H., "Group Sunspot Numbers: A New Solar Activity Reconstruction." *Solar Phys.* (1998) 179, 189

[271] The aamin data are at varying points in the diagrams to follow from Nevanlinna, H., Kataja, E., "An Extension of the Geomagnetic Index Series aa for Two Solar Cycles (1844-1868)." *Geophys. Res. Lett.*, (1993) 20, 2703

[272] The "Upper" Gleissberg can be counted as 72- 118 years. The "Lower" Gleissberg is counted as between 34-68 years.

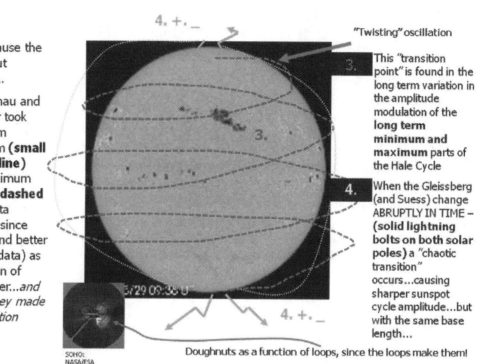

1. Loops cause the doughnut current...

2. ...so Duhau and de Jager took long term minimum (**small dotted line**) and maximum (**larger dashed line**) data (proxies since 1600 - and better present data) as a function of each other...*and found they made a "transition point"*

"Twisting" oscillation

3. This "transition point" is found in the long term variation in the amplitude modulation of the **long term minimum and maximum** parts of the Hale Cycle

4. When the Gleissberg (and Suess) change ABRUPTLY IN TIME – (**solid lightning bolts on both solar poles**) a "chaotic transition" occurs...causing sharper sunspot cycle amplitude...but with the same base length...

SOHO; NASA/ESA

Doughnuts as a function of loops, since the loops make them!

*Figure 1. The toroidal field ("doughnuts") up and down the Sun's axis is actually the maxima FUNCTION of the outward-looping poloidal field, or, the minima, that strikes a "current." The toroids thus powered create a "twisting" oscillatory motion up, down and outward off the solar axis. Examination of proxy and actual data reveal these two functions together make "**transition**" points in time probably from inertial, internal solar spin and resultant surface-oriented tachocline action. Longer-term cycles cause a sharper amount (amplitude) either for more, or for less, sunspots—but with the same base length.*

Solar maxima and the solar minima play out "normally" or "regularly" over time as we have seen. We observe them then over a relatively long (by human terms) time-expanse of "regular" solar minima and maxima in space, backwards into time as recorded by the savants and later government and institution scientists, till now. As such they have been numbered as cycles since the late 1800s (currently we are in Solar Cycle 24). Here, the search was for patterns relative to finding the strange and misunderstood "grander" phases of extended maxima and extended minima within the "normal" run of the data. The interesting thing here is the "dissection" of the actually "shorter-term" aspect of the Schwabe and the Hale Cycle, and, "together," their 22-year duration. Without going too far into what harmonics and quasi harmonics are for now we point out that the fascinating and still poorly-understood 22-year Hale Cycle consists of two "quasi-harmonic" beats. These "beats" translate into meaning non-linear and abrupt (or, potentially, chaotic) boundary shifts in the Sun. These two "quasis" are every-other-decade in length (from c. the Hale 22 year or "bi-decadal," or, 20 years) and spread further to every 46-60 years in length (or "semi-secular" [273]). These quasi-harmonic pauses occasionally intertwine closely, following each other at, on, or near, the transition points. This partly composes what we know of as a Gleissberg Cycle (of c. 80-88 year duration). Gleissberg Cycles, such as they are in this context, graph onto the c. 200 year De Vries Cycles like hybridizing two apple trees. What implications there are here for either are unknown. We now revive the diagram that was first shown in a less developed manner much earlier on in this book:

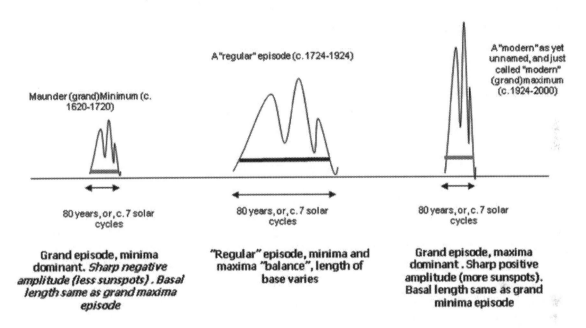

Figure 2. Negative, "regular" (normal) and positive (grand) oriented amplitudes in the solar tachocline due to chaotic transitions of aamin (poloidal loops) and Rmax (toroidal accordion-style movement).

The much-discussed basal lengths shown in Figure 2 consist of the quasi harmonics of the Hale and form part of the Gleissberg. But, the sizes of the amplitudes in all three contexts, from left to right (looking at Figure 2) covers—in Earth years relative to the phenomena discussed and their physical durations—around 200 years apiece (more like a De Vries Cycle) despite a driver of chaotic periods in the c. 80 year timeframe for both grand phases shown. The height in the "peak" of the grand minima episode (left, Figure 2) means less sunspots and less brightness (bearer of reduced or lower solar activity) and the height in the "peak" in the grand maxima phase (right, Figure 2) means more sunspots and more brightness (bearer of increased or higher solar activity). [274]

---

[273]   Or, half a century

[274]   The Gleissberg is split into two levels: Lower Gleissberg (34-68 yrs) and Upper Gleissberg (72- 118 yrs).

The one thing that is oddly the same for both grand minima and maxima is the base lengths of either. What is this, if not why is it? In the case of solar variability, the important events are those in which the dominant cycles change abruptly in time, which happens during a chaotic transition. This indicates that at the time of a transition, an abrupt change in some boundary condition occurs. What the change is, where in the solar boundary, is unknown.[275] The observed same-base lengths could indicate the action of the same physical actor in either case. We label the amplitude modulations of the sunspot cycle as a succession of "quasi-harmonic periods." These quasi harmonic periods during which the modulation of the sunspot cycle amplitude are a superposition of modes of oscillation have highly variable amplitudes, but nearly constant ("base") lengths. These oscillations are, then, the "quasi-harmonics" of our argument. The causes of amplitude disruption, which you could label as anti-harmonics, is purely internal, so far as the mathematics based on observations has it, to date. For solar-systemic influence on the Sun (the orbital force of for example major planets upon the Sun) has yet not been quantified and as such, is suspect as a driver of solar grand phases.

Planets do not seem to force the Sun regarding grand phases

The Gleissberg Cycle modulates the Hale Cycle's strength in its 22-year strophes and is probably a long-term inertial spin cycle. The Gleissberg Cycle seems to be vibrations in the inner solar core spin and out to the solar convective envelop (because of conservation of angular momentum) in the inertial reference coordinates. As such it probably has great bearing on both the Hale Cycle and the more "Earth-oriented" De Vries Cycle.

The Sun disturbs itself chaotically, internally (by itself) *three orders of magnitude greater* than external planets' force would do, as measured, from the data.[276] Hence the notion of how the Sun could be externally forced to cause such perturbations that may be construed as "chaotic," and to force transition points across regular minima and maxima, is still unmeasured. No data for this (or hard math-measured data, argued to the theory, against observation) exists yet. So the inertial-forcing of the Sun theory in the context of a barycenter shift—first proposed by Isaac Newton [277]—where our Sun moves inertially in the Galaxy every 179.8 years (and

---

[275] A good area to focus attention of research

[276] De Jager, C., Versteegh, G.J.M., "Do Planetary Motions Drive Solar Variability?" *Solar Physics* (2005) 229: 175-179: We examine the occasionally forwarded hypothesis that solar activity originates by planetary Newtonian attraction on the Sun. We do this by comparing three accelerations working on solar matter at the tachocline level: Those due to planetary tidal forces, to the motion of the Sun around the planetary system's centre of gravity, and the observed accelerations at that level. We find that the latter are by a factor of about 1000 larger than the former two and therefore cannot be caused by planetary attractions. We conclude that the cause of the dynamo is purely solar.

[277] "The Sun's barycentric motion was first demonstrated by Isaac Newton in 1687 in his *Principia Mathematica.* Newton (1687) showed that the sun is engaged in continual motion around the centre of mass of the solar system (i.e. the barycentre) as a result of the gravitational force exerted by the planets, especially

internally disturbing it due to the major planets' gravitational force orbiting around it) is as yet quantitatively unfounded.[278] Of course, that our Sun is inertially moved every 179.8 years is correct. The planets' orbits also realign after the shift. But whether or not this 179.8 year barycenter shift is a motive cause of putting the Sun into deep minima or high maxima phases in for instance chaotic transition points from external influences is for now, at least, not founded on computational fact. It has been well theorized. Armed with these measurements, telling us what the Gleissberg Cycle is in relation to the Hale and to other solar cycles, it is easy to look at Figure 1 again and to ponder the force generated by toroidal up, down, and outward-emitting power—stored in the tachocline—and being revealed in the longer-term Hale-Gleissberg-De Vries connection.

In this context, we consider the discussion carried over so far on the solar dynamo.

The partial Gleissberg Cycle (curve) as an internal spin cycle could perturb the Sun three orders of magnitude greater than external planets do. There is doubt then, as to whether or not external forces in the Solar System (at least) conspire to force the Sun into grand phases.

Jupiter and Saturn. He came to this conclusion analytically (not by observation) by working through the consequences of his law of universal gravitation." From page 3 of *Rhodes Fairbridge*, by Richard Mackey (Canberra), epitrochoid@hotmail.com

[278] My co-authored article, "Eighteen Hundred and Froze to Death," *Mercury* (Astronomical Society of the Pacific [ASP] San Francisco, Ca., USA) Sep-Oct 2003 outlines this in some depth and the data in it is, due to the findings here, are so far falsified insofar as they relate to external forcing of the Sun by major planets after a barycenter shift.

The transition point as sometimes chaotic and sometimes not: the motor of grand phases?

Longer term solar change prediction

Direct solar field strength measures only date from the early 20th Century. Solar polar field strength measures began in 1976. Proxy measure, then, must be used to fill the gaps of our understanding. The strength of the toroidal field component, Rmax, is based on the sunspot number "R" at normal solar maximum. As has been shown, this part of the puzzle has been fairly well recorded since 1610 and the days of Galileo, Christoph Scheiner, and others. The normal solar minimum component, "aa," is altogether another problem. As seen at the end of the last chapter, the geomagnetic index [279] aspect of solar activity is a constraining factor in overall understanding of solar variation: unfortunately in the longer term as well as in the even less-understood shorter term. The "aa index" as it is called, is assumed to be constant over time in its calibration. However, this is probably not the case and like all "constants" as regards solar activity, quite changeable and complex.[280] Additionally, due to the recording of the aa parameter over time and some Earth solar recording stations having been geographically shifted, the aa index might actually be somewhat lower than recorded in the more widely-used so-called standard values. These data were revaluated (as "Lockwood": 2006, 2009) and are shown side by side with the standard geomagnetic index data in this book. [281]

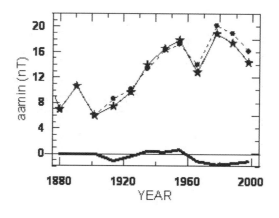

A comparison of the time series of the Standard aa data ((ftp://.ngfc.noaa.gov/STP/ SOLAR DATA/ REALTED _INDICES /AA_INDEX/AA_YEAR) (the dots) and the Lockwood data (the stars; private comm., 2009). The line at the bottom of the figure is the difference between the two time series. (nT means the Southward Interplanetary Magnetic Field)

---

[279] The longest time series for the geomagnetic index is from Mayaud, P.N., "Analysis of storm sudden commencements for the years 1868-1967." *Journal of Geophysical Research,* 80.

[280] Svalgaard, L., *et al,* "IHV, a new geomagnetic index," *Adv. Space Sci.,* 34, pp 436-439 (2004)

[281] Lockwood, M., *et al,* " The long term drift in geomagnetic activity: calibration of the aa index using data from a variety of magnetometer stations," (2006)

The ultimate value of the geomagnetic index, constant or inconstant, is its use as a suitable proxy measure of the sun's polar magnetic field flux at solar minimum. For it appears as if there is a fine linear relationship between "aamin" values and the observed polar field strength at solar maximum. This latter being called the Dipolar Maximum (DM). [282] This linear relationship is clearly shown in the plots shown below in Graph 1.

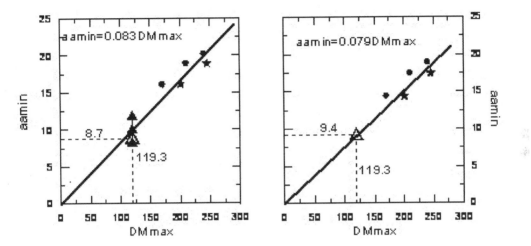

Graph 1. The geomagnetic index at solar minimum vs, observed maximum amplitude of the dipolar field strength DMmax. "Standard" data to the left: "Lockwood" data to the right.

That the tachocline, a thin, under-the-convection-layer dynamo "pump" is affected cyclically, is key to the theory as to why the Sun may go into a deep minimum or maximum phase at chaotic transition times and in and of itself demands falsification. This "maximum-generator" of accordion-like motion with its concurrent gas-particle-wave push, often involving much force indeed (perhaps received from solar-core vibrations) is the logical place to look for deeper maxima. Or, should a weakness factor be the case, it is a place to look for quite short maxima. There is, then, the absence of much solar internal perturbation if a chaotic transition for a sharper amplitude occurs (see far left in Figure 2). This causes a grand minima "type" chaotic transition.

A phase diagram such as that shown in Figure 3 can show a close-to-linear relationship between the two parts of the solar dynamo motor for maxima and minima. A phase diagram can show the relationship between maximum sunspot numbers and minimum geomagnetic "aa index" values.

---

[282]   De Jager, C., and Duhau, S., "Forecasting the parameters of sunspot cycle 24 and beyond," *J. Atm. Solar Terr. Phys.*, 71, 239 (2009)

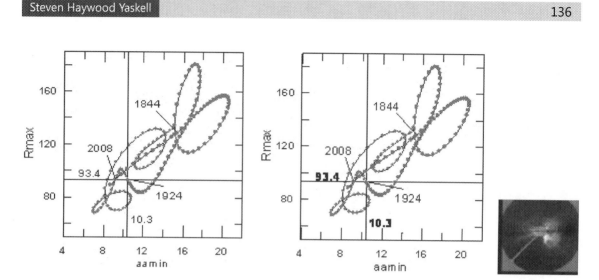

Figure 3. Partial "phase diagram" of Rmax (maxima) and aamin (minimum) data plotted from 1844 to 2011. Left hand plot is the standard data and the right hand plot is Lockwood (Data prior to 1968 is from Nevalinna and Kataja.[283]) Lower analemma-like dotted lines mean "normal" or "regular" phases and the higher ones indicate extended maximum conditions. Far right shows a reference diagram of the actual phenomenon as it "looks" occurring in Nature.

The most interesting part of Figure 3 is that is shows the crucial solar reference points 93.4 (poloidal aamin) and the 10.3 (toroidal Rmax) where potential grand phases have occurred (in 1924, and as posited in this book, once again in 2008.

The most interesting part of Figure 4 is a coordinate-point crossover when the (Rmax in Figure 1) toroidal field ("doughnuts") up and down the Sun's axis go against the outward-looping poloidal (aamin in Figure 1) field, and so "striking a current." Thus the toroids create a "twisting" oscillatory motion up and down the solar axis as Figure 1 attempts to describe. This oscillatory motion, twisting, *is* the transition point, made either chaotically or not chaotically, and which repeats itself in time, perhaps cyclically, as is shown briefly in Figure 4, and on a flat-plotted graph in Figure 5.

---

[283] Nevalinna, H., and Kataja, E., An extension of the geomagnetic index series aa for two solar cycles (1844-1868) *Geophysical. Research. Letters, 20,* 2703 (1993)

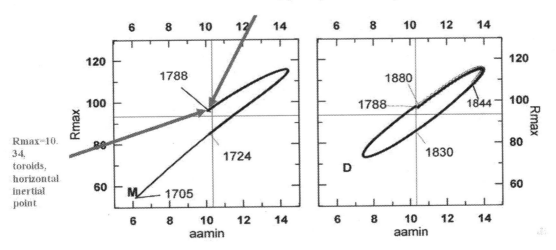

Figure 4. Complete "phase diagram" of the long term trend for the period 1705 to 1880.[284] Aamin—poloidal loops and Rmax toroidal rings from the tachocline cross the "transition point" (at 10.34, 93.38, inertial reference coordinates). The thin line in the left diagram corresponds to the ascending branch of the Maunder Minimum (M).(D is for the non-grand phase Dalton Minimum). The thick lines in the two figures correspond to the 1724-1924 "regular phase." In 1620, 1788, 1880, and 1924 the Gleissberg Cycle's path, the phase diagram shows, passed near the transition point's origin. In the case of 1620 and 1924, both were grand phases (one a grand minimum and one a grand maximum).

Proxy and more recently-collected solar data shown in Figures 4 and 5 reveal that the axes crossed by the poloids and toroids in longer time scales do so—or at least have done so—at specific coordinates, from inertial spin perturbations *relative* to the Gleissberg Cycle's quasi harmonic pauses.

Think of a woof and warp in a fabric-making machine: one (poloidal) going horizontal while the other (toroidal) goes up and down (vertical). The poloidal and toroidal fields overlap at the fabric-making point (chaotic) at the 10.34, 93.38 coordinates shown in Figures 4 and 5. There is reason to believe that this pattern will repeat in the future since it has apparently repeated itself in the past. As shown in Figure 5, the toroidal (Rmax) criss-crosses the physical action of the poloidal (aamin) field at 93.38 coordinates due to the action of the thin but powerful tachocline. The poloidal (aamin) accepts this crisscross at the 10.34 coordinates.

When the toroidal and poloidal action of the Sun somehow coincides with these two coordinate points after a spin-cycle shock, plotted with proxy and recent solar data, a transition occurs. Yet this transition is not always a chaotic transition. The chaotic transition depends, apparently,

---

[284]   The aamin data prior to 1844 are an extrapolation (ibid) from Nagovitsyn.

on how and what the spin cycle reacted at or to at these coordinate crossover times. We posit that it was from some internal cause, external Solar Systemic forcing events currently nugatory as regards extended solar phases (grand phases) as previously posited on strict mathematical grounds as they currently exist.

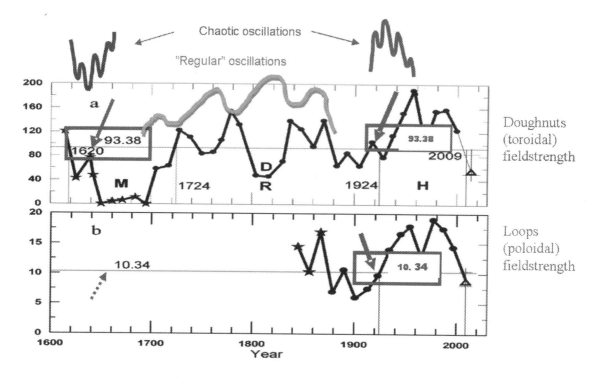

Figure 5 (a and b) (top a, bottom, b). Figure 5 (a) shows two kinds of grand phases stretching out across the latter end of the millennium using "best" data available. The Grand Minimum (M: 1620-1724—named the Maunder) and the Grand Maximum (H: 1924-2009—designated H) and the non-chaotic "Regular oscillations" (or, R: 1724-1924). Figure 5 (b) below it shows data for the recent leg (c. 1840 on) and showing the recent crossover between inertial spin coordinates from the poloidal [10.34] and toroidal [93.38] fields) and corrected, which helps the case for establishing the recent Grand Maximum phase which now is apparently over. Arrows show congruence with transition origin points.

Figure 6 elaborates on points made back in Figure 2. Figure 2 shows the base lengths in the "positive" aspects of a chaotic transition for either up or down and the "negative" regular (normal) period with a larger base draped in the middle. Figure 6 takes this data and plugs it into an actual historical solar dynamic.

Quasi harmonics: tracing of nonlinear dynamic motion in long-term ways

But now we look at the single amplitude loops shown in Figures 2 and 6 and widen them into two lines. These two lines, the quasi-harmonics of the torsional tachocline oscillation ("quasis") either are:

- Every-other-decade in length (from the Hale 22 year), or, c. 20 years (bi-decadal)
- Every 46-60 years in length (semi-secular)

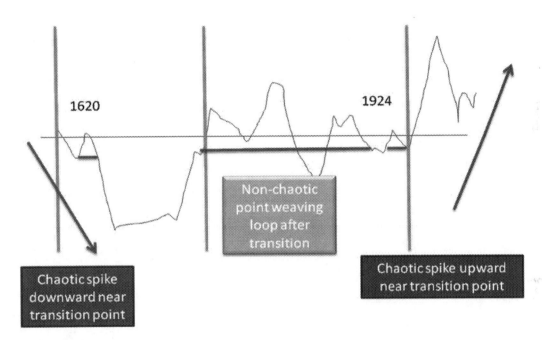

Figure 6. A simplified curve drawing of the events overlaying Figure 2. The chaotic downward spike leading to the grand solar (Maunder) minima and the chaotic upward spike leading to the recently-departed grand solar (Modern) maxima. The dynamic action for this probably arises from the tachocline driven internally (from core inertial spin) to a chaotic transition point. The period from c. 1720-1924 was, roughly speaking, "normal," or "regular" in that the transition point in 1724 at 10.34 and 93.38 was non-chaotic. It was characterised by a long basal structure and looser amplitude loops. That is, the "semi-secular" and "bi-decadal" (Figure 7) or "quasi-harmonics" did not overlap.

and sometimes, they follow each other closely, at times overlapping each other at, or near, tachocline-from-internal spin cycle-instigated transition points (Rmax and aamin: the usual maxima and minima in other words). This, then, is what would be called, in sum, the *actual* Gleissberg Cycle of c. 80-88 years. [285]

---

[285]   Between the minimum (lower limit) of 34 and maximum (upper limit) of 118 of the Gleissberg Cycle, you arrive at 84, or, the c. 80-year cyclic span.

In 1724 a sudden decrease in the amplitude of the Gleissberg Cycle occurred at the start of a "regular" (R, for example, in Figure 5a) or what we have been calling a "normal" phase. Earlier, in 1620 and again only much later, in 1924, the Gleissberg Cycle passed close to the origin of this tachocline/inertial spin cycle push and simultaneously, the semi-secular and the bi-decadal oscillations passed close to the origin of the tachocline inertial spin, as well. Recently it occurred yet again.

At these moments, the Gleissberg Cycle suddenly increased its lengths and amplitudes, apparently related to the grand episode of the M (Maunder)—the downward amplitude "spike," and the H (Grand Maximum)—or the upward amplitude "spike" (see Figures 2 and 6 far left and right and compared with Figures 5a/5b and 7). Once again, picture a car driver on a country road suddenly shifting gears into reverse from a forward oscillation spin on the car's differential. To the contrary, in 1788 and 1880, the semi-secular cycle was strong in spite of closeness to the point of origin. At these dates the Gleissberg Cycle continued its "weaker" amplitude cycle, which is a characteristic of a long "regular" or long "normal" phase or oscillation. The car's differential was shifted, but perhaps not so violently.

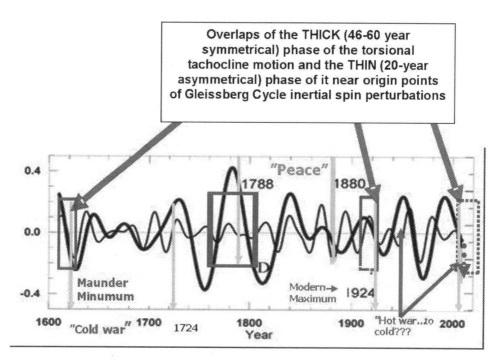

Figure 7. The parts of the torsional tachocline oscillations as seen in the semi-secular (thick line, each 46-60 years) and the bi-decadal (thin line, every c. 20 year) oscillations, respectively. In 1620, 1788, 1880 and 1924 the path of the Gleissberg Cycle passed near to the transition point origins (lines next to 1620, 1724, 1788, 1880,1924, 2009). The data here suggest a strong downward overlap of the two oscillations near 1620 and 1968 (see Figures 8 and 9).

In light of their interpretation of the solar dynamo, Duhau and de Jager conclude that "the torsional oscillations stabilize the tachocline-convective layer system motions, leading to transferral of angular momentum in the latitudinal (horizontal) direction from the core spin to the tachocline." This would account for alternately the lack or the abundance of sunspots in the mid-solar latititudinal range, respective of which kind of grand phase is chaotically being kicked into play (see Lorenz force from Chapter 6).

Building off the above information, Duhau and de Jager conclude it seems that "after a given transition, the amplitude and length of the Gleissberg Cycle depends strongly on the very phase of the torsional oscillations. This then determines the apparent random evolution of the solar dynamo."

Solar dynamo evolution is determined by a regular sequence of three well-defined quasi harmonic periods separated by very brief chaotic transitions. In Figure 7, we see exactly where an overlap of the two parts of the torsional tachocline oscillations—literally, the Sun's "twisting" motion, kind of like a rubber band being wound and unwound—in each 46-60 and 20 year "dance." Notice their strong overlaps in 1620, somewhat more so in 1924—and both of these heavy overlaps occurred at times when *chaotic* transitions occurred, resulting in *tighter* amplitudes—one tightly down (1620) and one tightly up (1924). (The transition points are light-colored arrows headed downward in Figure 7). Again, glance at Figure 6 as a reference point to see the total effect internally. (Figure 1 shows the total effect somewhat more externally as a cartoon.)

Observe also that in 1724 (in Figure 7) the overlap at the transition points with these two parameters (each 46-60 and 20 year oscillation) is much less pronounced. This transition was perhaps less if not at all chaotic: it resulted in around 200 years of "regular" ("normal") solar behavior reflected by solar proxy and cultural data in an Earth climate not especially "extreme," so to speak, compared to what was recorded in the Maunder Minimum (cold "never like this before") and in the recent Modern Maximum. This same "regular" phase could be approaching once again. But the last years of the Modern Maximum are still characterized by more noticeable Earth warming, especially from Greenhouse Gases like released or wandering Carbon Dioxide and abundant hovering water vapor. Note well in Figures 7 and 8 that the Dalton (D) fits comfortably in between one lazy downward 46-60 year loop, thus removing it from grand phase status compared to the extrema shown on either side. Grand minima and Dalton "type" minima are due to different aspects of the solar dynamo altogether. Grand episodes are due to a sudden change in the secular oscillation's length and amplitude. The Dalton was not due to this; it was the result of a semi-secular oscillation where the relative minimum occurred in synchrony with the relative minimum in the secular oscillation.

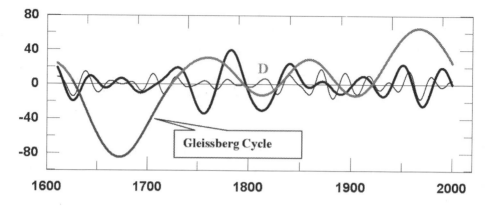

Figure 8. The three components in which the long term variation in R_{max} and aa_{min} as presented in the phase diagram in Figure 3 may be decomposed, as shown here for R_{max}. These are: the Gleissberg Cycle and the semi-secular and the bi-decadal oscillations (thick and thin black lines, respectively).

Present loomings into the future: regular or grand? Shorter term prediction of solar activity; among others, the Hallstatt Cycle

The dynamo system underwent a chaotic transition from the grand Modern Maximum of the 20th Century's 1924-epoch, this having come to a close in 2009 perhaps from solar action occurring in 1968. Figures 7 and 9 show this occurrence. The magnitude of the sunspot maxima happening at the end of a chaotic transition depends on the type of phase that develops afterwards.

The data interpreted in this section eventually approaches one cycle touched upon in Chapter 3—the multi-millennial Hallstatt of c. 2, 200-2,300 years (ironically, what we labelled a long term "sun" cycle). In this consideration we look at Solar Cycle 24 (begun in 2008) for formulating a tool for improved sunspot count prediction. The variability of the relative phases and strengths of the bi-decadal and semi-secular oscillations at the start of the forthcoming (*if* forthcoming) grand episode are still not known. Thus it is difficult to evaluate the precise sunspot number value of the maximum for Cycle 24. The question at hand is, will it be the start of a grand episode or a return to "regular" oscillations?

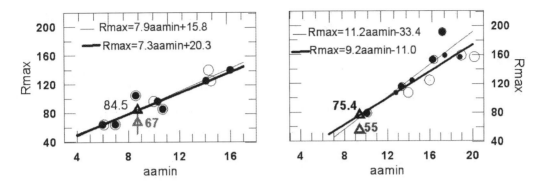

Graph 2. Sunspot maximum data Rmax vs. the geomagnetic index data aamin. Either panel left to right represents Regular Phase (Sunspot Cycles 9 to 25) and Modern Maximum (grand phase) sunspot cycles (Sunspot Cycles 16 to 23)

Key to Graph 2

- Open circles, thin regression line = geomagnetic aamin data
- Filled circles, thick regression line = Lockwood's time series geomagnetic aamin data
- Values plugged in (abscise) as 67 (left) and 55 (right) upward triangles are from standard and Lockwood data respectively.
- Value plugged (ordinate) as 84.5, respectively 75.4, correspond with those of the geomagnetic index at solar minimum vs, observed maximum amplitude of the dipolar field strength DMmax (see Graph 1).

Earlier forecasts predicted a sunspot number of 68 +/-17, or about 55 using standard as well as Lockwood aa data and this is still traceable in Figure 5a and 5b. Since a value exists for present aamin data in Sunspot Cycle 24 based on observations, the nonlinear relation between previous aamin data and the attendant Rmax value is used to improve predictions.

In Graph 2 the sloping aspect to the regress lines are formed by a relation between the respective secular oscillations. You could say that the relation is almost linear, yet the slope changes widely after the each grand phase transition. The secular oscillations as regards amplitude and length were similar in the grand episode Maunder and the grand phase called the Modern Maximum.

Due to the nonlinear relationship between bi-decadal and semi secular oscillations (the quasi harmonics) all points around the regression lines in Graph 2 are dispersed and displaced as regards each other. When the semi-secular oscillations are the strongest (as was the case in Sunspot Cycle 19, starting in 1944), the dots or points in turn deviate the most from the line. (See the uppermost black dot in the right hand panel of Graph 2.) The maximum between

Sunspot Cycle 19 and 20 occurred in 1951. [286] That was when the semi-secular oscillation was positive (upward thick line for the year 1951 shown in Figure 8.). At the time of maximum for Sunspot Cycle 24 the semi-secular oscillation will be negative (thick line downward): this explains why the predicted values in Graph 2 (next to triangles) fall below the regress lines.

So given this data, it could be that a weaker semi-secular (negative) loop means the start of a grand episode (that is, Maunder-like) or hopefully, merely a return to more regular oscillations that were the norm before 1924. But since the variability of the phases and the strengths of both the bi-decadal and semi-secular oscillations at a coming of a grand phase are not known, predicting a maximum sunspot number for a sunspot cycle is imprecise.

Arrived at is a number equalling a 62 +/-12 sunspot number, not larger than 74 but not smaller than 50. This applies to either a regular episode having been brought in by Sunspot Cycle 24 or a grand episode (that is, a deep minimum). This is arrived at since any point will fall close to the regress line in Graph 2 when both the semi-secular and bi-decadal oscillation are small. The black triangles in Graph 2 should be close to the real value of Sunspot Cycle 24 if the two shorter oscillations have a small amplitude. Figure 8 shows the extrapolation that both oscillations will be negative in Sunspot Cycle 24. Each amplitude (upward loop) should consist of from 0 to -5 and -10 to -20 in total sunspot number.

A downward twist to the plot? The thin likelihood of a descent into a new maunder minimum

Figures 9 and 10, when studied against Figure 3's phase diagrams, seems to describe the continuous path of a cracked whip as it flings itself out into time: the 1788 and 1880 leg of the thick-lined Gleissberg Cycle quasi results, in solar terms, in a reduced chaotic effect in the 1924 transition. But that all could change, as the transition point is seeming to result in a lessening in the slack of the whip-snap. To use a more appropriate metaphor, the lines are behaving like water in a sharply-tilted hose that is alternately turned up and then, after a period of flooding, reduced by inertia. Then it heads down. But Figure 10 is only a worse-case scenario and nothing more (that is, it is factually baseless at this time).

As old texts [287] on electricity used to describe, the fluid flow of water in a hose, so constrained in a Venturi-like tube, is exactly how magnetism behaves, as well.

Other researchers have found corroborating evidence for a downward trend in Schwabe Cycles (with the attendant 22-year magnetic Hale) and the "200-year" cycles (quasi-Gleissberg), such as Abdussamatov (see Figure 11). This leads to smoking gun-like thoughts that a grand episode is immanent. The 11-year Schwabe cyclic variations of TSI occur in relation to components of

---

[286]   See graphs for Sunspot Cycles 12-24 in Chapter 8 for another view of this data.

[287]   "The Story of Electricity and Magnetism," in *Popular Science Library*, ed. by Robin Beach (Collier:1938)

the 200 year cycle, as Duhau and de Jager have taken pains to show. What Figure 11 shows is an uninterrupted series of the TSI flux density from 1978 (bold line on graph) as measured directly by space-based (not Earth-based) instruments. The smoothed 11-year (Schwabe) of TSI at the maximum period of the 200 year cycle ("quasi"—Gleissberg) was nearly equal to 1.0 W/m2 or 0.07%. This has been on the decline (the Schwabe) since the early 1990s. Cross-correlation and quasi-parallel alignment curves of 11 year variations of solar activity and of TSI, both in phase and in amplitude, were made. [288] The 200 year component of TSI variation showed a quick decrease (dotted line in Figure 11) since Solar Cycle 21 through Solar Cycles 22 and 23. Taken together, the parallel-occurring decrease in the 200 year cycle component of the sunspot activity (Schwabe-Hale) and the TSI may reveal an on-going descent of the 1924-initatied grand solar maximum.[289]

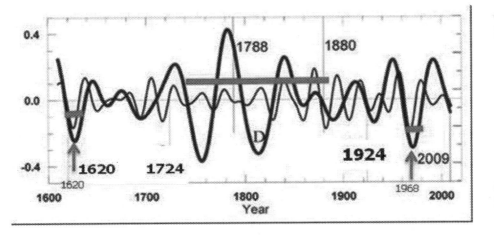

Figure 9. Downward loop "overlaps" of the 46-60/ c. 20 year oscillations corresponding to near-transition origin point. The data show the overlaps in 46-60/c.20 years symmetries in the transition point to be oddly similar: the one occurring around 1968 at mid-downward point in the axis is especially deep with a sharp downward spike—much like the 1620 apparition (arrows, bottom, left and right). The amplitudes gained are remarkably similar—very unlike the W-shaped loops spanned by the long line in "normal" times, the symmetrical and asymmetrical loops being way out of "synch."

---

[288]   Abdussamatov, H. I., *Kinematics and Physics of Celestial Bodies* (KphCB). 2005, 21, 471

[289]   Abdussamatov, H.I., "Russian project Astrometria to measure temporary variations of shape and diameter, TSI etc of the Sun" (2009), p. 3. "The amount of solar energy supplied to the Earth is directly linked to the value of the solar radius, in the other words to the radiating area of our star. Cyclic variations of the TSI occur due to the oscillations of solar radius with amplitude up to 250 km within a "short" 11 year cycle and up to 750 km within a Grand 2-century cycle." Calculations showing this using Stephan-Boltzmann equations are on page 5 (of Abdussamatov)

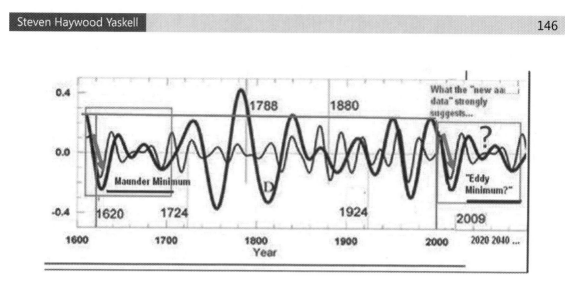

Figure 10. The coming Eddy Minimum and the Devil's Advocate view. Based on data shown in the phase diagrams and other sine curve diagrams in this chapter, a guess is taken as to the severity of the coming deeper minima (far right).

Improving the prediction?

Why was the new prediction of 62 +/-12 sunspot numbers, not larger than 74 but not smaller than 50 valid at all? Figure 12 shows the plotting of two consecutive sunspot cycles similar to Sunspot Cycles 23 and Sunspot Cycle 24.

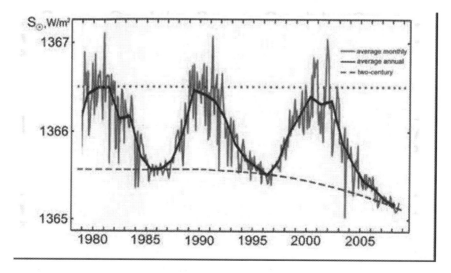

Figure 11. Eleven-year and two-century cyclic variations of TSI for the period of November 16, 1978-November 29, 2008 (Frohlich C., Lean J. Astronom. Astrophys. Review. 2004, 12, 273; www.pmodwrc.ch/pmod.php?topic=tsi/composite/SolarConstant ; Abdussamatov H.I.)

In Figure 12, the upper panel corresponds to the cycle pair occurring near the transition phase to the Maunder Minimum. The bottom panel shows Solar Cycles 11 and 12 (roughly A.D> 1867-1878) this latter panel being ensconced in the "regular" oscillation phase from c. 1724-1924. The strengths of the second cycles in each panel pair are comparable to the expected strengths of Sunspot Cycle 24.

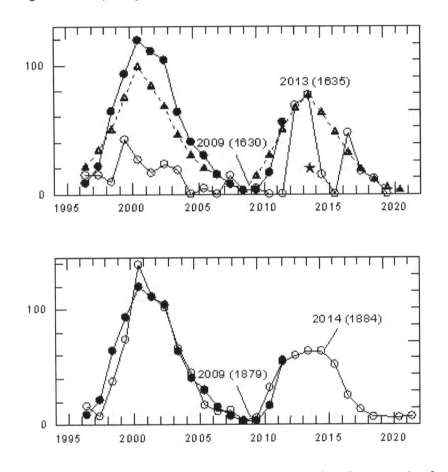

Figure 12. Improved prediction of coming Sunspot Cycle 24's outcome based on two pairs of consecutive sunspot cycles.

Figure 12 is the updated if not improved prediction of what Sunspot Cycle 24 will yield. The estimated error is the vertical line from 0 to 100. The filled and open circles are taken from data shown in Figure 13 regarding Schwabe Cycles since c. A.D. 1600. However, they have been shifted ahead by 378 and 130 years for the upper and lower panels in Figure 12, respectively.

Figure 12's upper panel includes Nagovitsin's (2006) [290] annual sunspot number time series shifted forward 378 years and Vaquero et al (2010)[291] data (the black star).

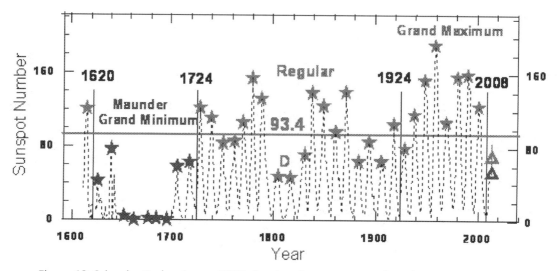

Figure 13. Schwabe Cycles since c. 1600 showing the sunspot number plotted against time

In 1629 and in 2008 (upper panel Figure 12) the was a transition to the actual Maunder Minimum and to a potential forthcoming grand episode (starting in 2008). By contrast the year 1879 (Sunspot Cycle 12's start) was well within the 1724-1924 Regular Oscillation period shown in Figure 12's bottom panel.

Sunspot Cycle 24 will mark the first cycle since Sunspot Cycle 16 [292] (which started in 1913) with a maximum sunspot number below the transition point value. This was ostensibly before the start of the so-called Modern Maximum grand episode of 1924 which began its trek at the start of Sunspot Cycle 17 (1923) but did not show its increased output until Sunspot Cycle 19 and onward into the 1960s and 1970s.

Sunspot Cycle 24 will also be either the start of a new grand minimum or the start of a return to "regular" phases as described between 1724-1924 (see Figures 5a and 5b and Figure 7, mid-section oscillations where the gaps between the bi-decadal and semi-secular oscillations are wide and loose). Given the imprecise knowledge of the values for aamin (geomagnetic

290   Nagovitsyn, Yu. A., Solar and geomagnetic activity on a long time scale: reconstructions and possibilities for predictions. *Astron. Lett.* 32, No. 5, p 344 (2006). (Translation from *Pis'ma Astronomicheski's Zhurnal* 32, No. 5, 382.)

291   Vaquero, J.M., et al Revisited sunspot datta: a new scenario for the onset of the Maunder Minimum, *Astrophys. J.*, 713, 124 (2010)

292   See diagrams for all sunspot cycles since 12 in Chapter 8.

index value) and the polar field strength parameter (DMmax) for the preceding century, it is impossible to determine which of these two types of oscillations will devolve, using phase diagrams. It is most certainly however a descent downward from grand maximum phasing as witnessed in the latter half of the 20th Century.

The Hallstatt Cycle's significance in glimpsing the character of any coming grand minimum

An answer to the conundrum raised in the preceding section as to either a return to a grand phase or a regular one might be found in the Hallstatt Cycle. It adds to the evidence as to why any deep minimum episodes will be a thing of the past for many centuries to come.

An introduction to the Hallstatt in terms of long-term solar climate influences is found in Chapter 3 of this book. As was there described, a so-called "positive" aspect of this phase began in 1935. Since the Hallstatt Cycle is a "true" Sun-earth cycle, connecting the I's of Earth activity to the dots of long-term solar cyclicity, it seems as if the "negative" aspects of the Hallstatt could be tamped out for the millennium.

The last positive phase of the Hallstatt ended in c. A.D. 918 and began once again in A.D. 1935. This lapse from the Middle Ages to the ages of jazz is consistent with the numbers of 2,250-2,300 as given either by Steinhilber or Cliverd, also known as the length of the long-term Hallstatt Cycle. As noted in Chapter 3, the catastrophic climate events of 68 years prior to this descent into negative phase at A.D. 919 perhaps heralded severe global cooling hemispherically. Figure 14 also shows that maybe it was also the sign of the coming of at least four major solar minima (the Oort, Wolf, Sporer and Maunder). Solar and global events since roughly 1940 have indeed given credence to the idea that warmer conditions overall will be the case for another millennium. Given the distinctions made between what "Dalton-type" minima are vis Á vis grand episode minima as outlined above, Figure 14 in relation to the discussion in Chapter 3 on the Hallstatt Cycle makes even more sense. It is also consistent with the cultural/climatological slide-window data by Howell in that chapter, and could be a reason to why there will be no deep solar minima for another plus-1,000 years.

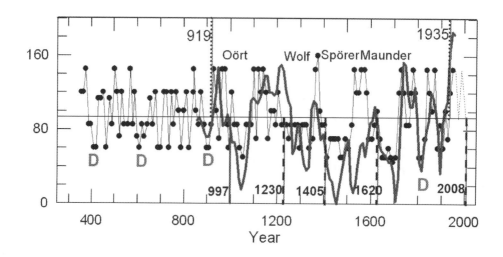

Figure 14. Two data series [293] recreating the path of the Hallstatt Cycle from c. A.D. 250 to the near future. Faint dashed line in the near future is Schove's prediction of sunspot maxima for cycles 19 to 24. The vertical dashed lines indicate transitions to grand minima (reviewed by Duhau and de Jager from Nagovitsyn: 2007). Vertical dotted lines hem in the recently-ended negative Hallstatt phase. D marks Dalton-type minima. Black horizontal line in the middle of the graph is the transition point level.

But so long as the sunspot maximum value of Sunspot Cycle 24 is not known, then safely predicting that no extended solar minima like the Maunder not occurring for another 1,000 years is impossible.

---

[293]   Solid black line shows Usoskin, I.G., *et al,* "Millenium-scale sunspot number reconstruction; evidence for an unusually active sun since the 1940s," *Phys. Rev. Lett.,* 91, No. 21, 211101-1. Dots are Ibid Schove (1955). The data are qualitatively similar, Usoskin's data more clearly outlining the Hallstatt Cycle.

8. A summary, some observations, and some closing thoughts

This book was not a book of answers. It was a book of questions. But perhaps the human mind must always be driven forward in rough pursuit of hard answers lest as Kepler [294] once implied, the borders of all things become too well known.

It is hoped that the disinterested enquirer with a knack for probing these stabs at calculated speculation can build here. In any case, this slap-dash house of fact, facts waiting to be discovered, and quasi-fact is an edifice that is most welcome to falsification on all or on as many points as is possible since this is a science book. Nothing in it is sacred. The additional observation is here noted that some of these facts are factoids boardering on urban legend and are entirely wrong. Do not excuse any unintended flights of fancy or downright error or both.

We begin with the Sun and its aspect. It is a journey of eight light minutes in distance and someday it could be bypassed with the help of technology in actual Earth minutes. But for now we muse upon its still truly great physical distance, the Earth like most other planets being blasted by the contents of Sol's heliosphere unevenly if constantly. Earth is warped and alternately relatively left alone by its star's magnetic / hydrodynamic force field. Nor is this sun of ours operating in some void. It too is prone to intergalactic and extragalactic forces that may add to its own force in the form, as we have seen, of galactic cosmic rays as a contributing power in considering MHD especially. If the Sun's very variable flux is reduced, these add to the cosmic ray punch a planet like Earth may receive or if strong, deflect these same rays. The Sun's pendulum-like shifting roughly every 180 years is also proof that it too is constrained and yet moving, and this so far seems to have nothing to do with the energy powering its varying cyclicity. It is internally disturbed it seems, to effect these ends, many magnitudes more highly. Ultimately Sol has no special place in the cosmos physically. We do not take up the matter of its spiritual significance here though that, too, is important. It heats us and protects us; perhaps has even been the direct cause of us. But it can unintentionally ruin as much as nurture.

Perhaps the most disturbing aspect in this discussion is that realm where the Earth's atmosphere, ends, and what we would consider to be space, begins. How particles move in magnetic whirlpools to effect electronics disruption and influence climate among other things begins here and how exactly in some cases is a task for future scientists. Perhaps these scientists will be born only in some distant future. The scenario wished to be formed here at the world's atmospheric top is what could be construed as "layers" of hydrodynamic/magnetic force and

---

[294] "The human mind shall never be lacking in fresh nourishment."

possibly many other influences besides. To reiterate, the word "layer" itself in this context is unfortunately a label and a miserable one besides: it suits neither the Sun nor Earth as a proper physical description. For one is instantly reminded of Aristotle's spherical delineation of distinct zones. This as in the perfect-spheres concept such a mental construct was once meant to convey and which is still a building block to higher thought. The Sun, furthermore, is a magnetohydrodynamically-driven ball of gas. It has nothing humans could construe as a "surface" at all. Some of the phenomena ranging off its "surface" pitches out millions of miles into surrounding space while phenomena right besides it is drawn inward.

This seething, writhing near-million mile in diameter magnetohydrodynamic, probably nuclear-fueled G-class variable star has as we have noticed some reasonably predictable behaviors and some regular, identified emitters. Perhaps the most predictable of all indices on Sol is its pattern of sunspots and the somewhat regular manner in which they appear and move through the star, and the Schwabe/Hale Cycles. The other cycles mentioned such as the Gleissberg and the more-Earth tied De Vries are much more suspect as regards notable regularity and even what they are, let alone where they are. The sunspots are themselves weak emitters and this is more or less a common thing: rarely do these "signs" of increased / alternately decreased (by sunspots' absence) solar activity shoot forth much Gauss. Yet for all this predictability and regularity there are grave and outstanding variations in behavior, quantities, time limits and other variables concerning even these relatively well-known bearers. Plages, faculae, and other emission zones have their measureable and strong radiation signals: prominences, plumes and still other phenomena that are as yet unknown.

The strongest emitters of radiation and particles are the solar flare and its SEP-punch, tied as it is at times to the CME and coronal holes. These phemonema, working in tandem, deliver the high-speed magnetised plasma (from the CME) and SEP particles (from flares) even with coronal hole action. This all in some high-velocity mushrooming rush which shoots into the solar corona and, forced outward by the solar wind at extreme heat, casts nonlinearly out into space all around Sol. It is this magnetic/radiation/particle mix that often touches our magnetically-enclosed atmosphere, and sometimes pushes it all the way back, ionizing it in such a way we term the ionosphere (and which is no sphere at all, remember). We have noted the arcane and still poorly understood types of radiation familiar in their wavelength sheaths from high school science class: the mysterious and suspect infrared (IR), the X ray, gamma (Γ), ultraviolet (or UV) microwave and others. Yet it is believed that particle interference from the Sun, coupling with UV / EUV in the stratosphere, is—unlike just with magnetic storms on Earth and other destructive interference in Earth's near space—most likely a co-modulator of Earth's own climate; Earth climate itself being perhaps a series of regionally-driven and not uniformly global, phenomena, working in tandem. For all this focus on natural climate modulation by the Sun, it by no means is meant to cancel out what local power over our climate Man himself can exert—most specifically on a local, regional basis.

What happens when solar flare / CME mixes strike the Earth, there to cause a "bow shock" to Earth? First we go "sphere by sphere" To summarize we return to Chapter 4 and consider the magnetohydrodynamic solar "force field" and its relation to Earth's magnetosheath. A few thousand miles up, and we see that there are nearly 5,300 miles of "exosphere" (*not a sphere*) that may be totally solar (and partly extra-solar or, galactically) steered. That is, this distance, *per se*, may shrink if the magnetohydrodynamic force exerted by the Sun or energy rods from Galatic Ray Bursters (GRBs) shortens the electromagnetic force field of Earth, and warps and woofs any wavelength energy emissions that are creeping into Earth or straining—or racing—to go past it, or to bounce harmlessly off it, if weak. Alternately when the "force" from the Sun (and the odd weaker GRB or two, together) weakens, the "distance" does "loosen up" and again, it widens back far outward. Hence, this part of Earth's "atmosphere" is conceivably almost entirely guided by the Sun-or extra-galactically (or both). That it is so is no cause for fear: it is perhaps due to this very arrangement in its peculiarity contra other planets that we or any life exists at all.

The next "sphere" down or, the thermosphere, close to 700 miles above the Earth's surface at its variable and magnetically-disturbed top (picture cloud tops and their variability at height as if from a jet aircraft) is also so affected. Do these zones that far up actually constitute anything we could call an Earth "atmosphere?" Is it an atmosphere of Earth?

What is, then, the definition of our atmosphere, if this is not so? In an age of redifinition, we wish to know.

At this point the non-spherical ionosphere begins, allowing that particle flux to transpire spinning or using some other motion or series of motions around and downward into the stratosphere at the 31-mile mark from Earth's surface. (The particle flux is often gratis an archimedian spiral driven initially by a solar-seated CME.) Could this then perhaps be the better beginning of what we term the upper Earth atmosphere? I hazard this, since the UV band of solar radiation is modulated by the ionosphere and at the D layer, in the mesosphere, we can see the Aurora borealis and related phenomena from our streets. Where it was hotter higher up, the temperature *decreases* in the mesosphere. Then we have a hotter place *below* the mesosphere, in the stratosphere (temperature rise with height) and in the troposphere below it, a temperature *decrease* with height.

The entire stratospheric-tropospheric mix is an electrified place and as even Gilbert knew, a magnetic one too, and has even been called "electrosphere" in a seemingly never-ending aerospace engineer-like way of framing difficult-to-understand physical entities who rely often on the terminology of academe. It is in the crucial areas of the ionosphere's D layer and the stratosphere where dynamic coupling of particles (to include the much-talked about SEPS) takes place, and which has that enormous effect on the Earth if it is somehow, at times, delivered as a weighty punch at Earth's very surface, right through the troposphere and into and under Earth's very lithosphere, crust, oceans, and perhaps even mantle. Modulated in this zone are the anomalous EUV particles thought to steer Earth's climate in and around the lithosphere. In other words, right where we live.

We can assume, given all the various goings on in the entire atmosphere—wherever we wish to begin it—and however so considered or layered, that it is anything *but* orderly up there, sitting in gentle equipoise amidst our dearest ideal visions of perfection, or even equilbrium, or some other stable posit in the mindset of how we want to see it. Yet tend to an equilbrium it does, if in a punctuated manner. There is no perfection and no regular orderliness that we discern. That is not to say there is none at all for any length of time. As all things tend to an equilbrium or a balance, so far as we know in the three-dimensional / three-state change universe, these distances of breadths of the "layers" do, however, tend to certain heights from the surface in a variability formulation that must contain predictable entities and identifiable limits, as sizes of them have been basically established. How they vary given the nature of the Sun-Earth connection is often a matter of pure conjecture. But vary they do. What is very noteworthy is the lack of free-floating particles at the topmost levels of the atmosphere, increasing enormously as it falls headlong into Earth's lowest atmospheric domain, where particles abound with thick molecules and dominate the friction-filled, storm-charged lower atmosphere levels.

But, more speculation about what we do not know follows. Given also the aforementioned "bow shocking" of Earth that accompanies a solar flare/CME combined punch from the Sun, what of the powerful vortices of something like the El Nino Southern Oscillation (ENSO) and other such geophysical phenomena? For example, the Hadley Cell (revisit the long if lively Chapter 3). Or any of the other manifestations of lower Earth climate, from the lower stratosphere to the surface of the oceans and land? Now in the same vein, though not directly connected perhaps, we recall for purposes purely of instruction the two kinds of motions on the Sun that operate in tandem that leads to such bow shocking on Earth in the first place. One was the thin but widespread tachocline, which is extremely flexible and most likely a bearer of the "regular" solar minima and maxima "machines" and the deeply-inserted, inertially-moved core which is most likely part of the Gleissberg Cycle, affected by the more predictable and better-known Hale.

It is perhaps this inner core cycle, self-disturbed in some nuclear fury in its various transitions, some short, some long—but maybe always of the same base length—which conspires to twist this tachocline "regular minimum and maximum-bearing machine" into oscillations that alternately widen or shorten a "normal" solar maximum . . . likewise to shorten any "normal" solar minimum. Or these can work to lengthen either. How much or for how long is still a mystery. This motion can perhaps shake a nuclear frenzy of solar activity on either of these dynamics into being that lasts for not only years in some cases but for decades: maybe even for centuries. In the case of the Maunder Minimum, the extended solar minimum did overlap two centuries. So, this internal spin cycle can be the sign of a nuclear fizzle, spinning the solar disk faster in one hemisphere if reducing most if not all electromagnetic emitta, then widen and make more oblate the sun, yet ultimately causing the star to be less bright as a sign of its lessened nuclear activity. Or it can do the reverse, bringing into play extended solar maximums as reconstructed by proxies such as the Holocene Maximum, the Medieval Maximum, and the Modern Maximum some say we have just witnessed. If there has been a recent "Modern Maximum" the data collected on it should be made immediately clear, as it need not be reconstructed with isotopes (as proxies) alone.

A clue to a stalled gear in the "celestial machinery" of the Sun has recently been noted by Frank Hill (see the end of Chapter 6 and the figure on the helioseismographic interior of the Sun). The discussion there concerned a noted interruption in the equatorward rush of the solar jet stream; the usual sign of the impending sunspot cycle (as in Sporer's or Carrington's Law): not the EARs and polar solar spots but the regular cycle at the mid-solar latitudes that is the Schwabe Cycle.[295] As the British business journal *The Economist* reported for June 16, 2011 related:

Frank Hill [US National Solar Observatory] and his team were the discoverers, 15 [296] years ago, of an east-west jet stream in the sun. They also worked out that the latitude of this wind is related to the sunspot cycle. At the beginning of a cycle the jet stream is found, like sunspots, in mid-latitudes. As the cycle progresses, it follows the spots towards the equator . . . Intriguingly, however, Dr Hill's studies indicate that the jet stream of a new cycle starts to form years before the sunspot pattern. This time, that has not happened. History suggests a new cycle should begin in 2019. If the sun were behaving itself, Dr Hill's team would have seen signs of a new jet stream in 2008 or 2009. They did not. Nor are there indications of one even now.

In 2012 in an interview for EarthSky online [297] Dr. Hill added this note to the observation that lead to a recent American Astronomical Association announcement that a prolonged sunspot absence is on the way (and the attendant magnetic calm?):

(Hill): This jet stream typically first appears at a high latitude on the sun, near the solar poles, approximately 10 to 12 years before the start of the solar cycle. It then moves first toward the poles, and then another branch appears. It moves towards the equator. We should be seeing the poleward branch of this flow for Cycle 25. That's the next cycle of sunspots after the the one that we are in, which is Cycle 24. We should have seen that flow back in 2008, and we still have not seen it. And so this leads us to believe that that there is something different about Cycle 25 than we have previously seen . . . When the sun has sunspots on it, the sun is a little bit brighter than when there are not. So if there's a lack of sunspots, then the sun is a little bit dimmer. By a little bit, I mean one tenth of one percent. It's a very small fraction.[298]

---

295   See Chapter 5 and Figure 1 as a refresher to this discussion, then reconsider the final pages of Chapter 6.

296   Ibid Hill, F., *et al* (*Science* May, 1996)

297   Ibid, Salazar (April 12, 2012)

298   This small fraction could be the bearer-sign of reduced center of activity emission across all of them: for instance plages, faculae, even CMEs and solar flares. This would translate, then, into increased "magnetic

(Interviewer): What does all this mean?

(Hill): It could mean a range of things. It could mean that the next sunspot cycle, after the current one, could be delayed by two to five years, at least, and perhaps longer.

Given what we read in the preceding chapters, is this an early calling card of the most strongly reduced Schwabe Cycle yet in a reduced Schwabe Cycle series over the last 40 plus years (as noted in Chapter 7)? As the Hale Cycle—construed by some as the Schwabe's magnetic "component"—is the bearing sign of a lot if perhaps not every single bit of magnetic/electromagnetic activity on the Sun, could this be a sign that the Sun is about to begin its descent into a "calm" that could be, as de Jager poses, either a return to a more regular solar activity pattern or—more ominously—the beginning of a deep minimum? Let us hope the Hallstatt Cycle, having entered a positive phase in 1935, denies us such dubious pleasure.

So little is really known about how our sun functions that confidently projecting a grand solar episode in the near future-to-mid, end of century is as fraught with unknowns as climate model statisticians confidently predicting CO2 warming catastrophe by the end of the same. Even if we could predict the episode within 50% of error, there would still be the attendant question: "for how long?" Ninety years is based on the proof of how long the 17th-18th Century Maunder Minimum lasted. It is not an exact prediction. "For how long" is even more of an unknown than the 50% prediction of guaranteed occurrence of a prolonged minima complete with all accompanying solar physics and hard math proofs (of which there are none). Chapters 6 and 7 should amply have underlined the score of this sad symphony.

The vital aspect of clouds and climate change: two views

That the weakened Schwabe Cycles seem to be a calling-card of a halt to, if not a severe weakenng of, a new solar cycle has begun a new journalistic regime. Yet there is still no proof that say, Cycle 25 will not start at all for a long period. It might just start so slowly that it never seems to get started, somewhat like the current cycle slowly started in 2008—only the next being perhaps even more drawn out than even Cycle 24 (the current cycle) was. De Jager's calculated guess, given his data and explanation, is that it will be a return to "regular" solar activity that resembled that occuring before 1924. Maybe a sharp drop off into a deep minimum is apparent. No one can truly say. Will it be death by ice, afterall, and not by fire? Interesting journalism and political activism on the matter is already ongoing

Such dramatic contrasts are pundit drama spiced with likely statistics that catch an ear and drum a beat but little else. The real phenomena and occurrences are very natural if not as orderly as we would like. The overall consilience to be derived from these if/then social questions could come in as undramatic a fashion as was ever imagined. Introducing a cooled balance into this

---

calm." (Recall that sunspots themselves do not emit much.)

fevered debate is difficult given the honor of the questioners concerned, their investments of time, stakes on reputation and grants whatever their titles or educations. But the honor of broaching such questions is not in doubt. It is probably the most laudable congruence of intellectual power at this century's beginning for confronting challenges that face us ahead as set out in this book's introduction. To get past the Sun we must first know it more thoroughly. Knowing its effects on Earth is part and parcel to knowing how the Sun functions. Knowing how it functions and when it is safe or not means we have mastered the Solar System and at least one star.

In the concern over cloud cover, for example (which is just one of several albedo-contributors to climate modulation under the 10-mile range of the atmosphere, as discussed earlier) the concern that Greenhouse Gases are building this up has met head-on with the CERN experiment's results (see Chapter 4, Kirkby *et al*) claiming that increased cloud cover on Earth actually is formed by cosmic rays due to a weakened Earth magnetic shield in space—which, consequently, happens to be the sign of a weakened sun that is letting "them," "in." A weakened sun, in fact, lets any powerful rays from exploding stars in our Galaxy go right past the Sun and—if they happen to be in the most advantageous direction—pile straight on into the Earth. The magnetosheath can only keep so many of these out in its flabby state. Hence, more cosmic rays. Thus, more cloud cover. This could block out all or much of the Sun's rays over time,[299] leading to global hemispheric cooling: an unlikely event in the short term. The experiment is fact: the proof of it actually happening in space or the lower Earth atmosphere on the other hand has yet to be seen. Compared to pure climate modeling at least it was a Stanley-Miller [300] type experiment of apparent likelihood.

Geo (or, global) climate models see Greenhouse Gas's total effect to climate change, mainly by humans (CO_2 and water vapor—two of the most prominent Greenhouse Gases: there are others. Industrial processes in concentration yield a great deal of the former). Clouds are hardly seen as a factor in the models. That clouds are seen by geo-climate modelers as adding to surface warming is a different thing. Clouds, the modelers outside their models contend, absorb long wavelength radiation (or, IR) from the surface of the Earth and radiate some of it back down to the surface yielding the chance of increased surface warming of Earth. In addition to this absorption and re-radiation of infrared radiation from the Earth's surface it is just reflected at Earth's surface. IR can intensify warming locally and exacerbate the Greenhouse Effect where it would be currently occurring (say along coastal Northeastern USA in July). Notably, any longwave radiation (IR) from the Sun needs clear skies to do this to the surface and the amount of solar IR is still considered to be small (most radiation from the

---

[299] Probably very deep time of such bombardment without relent is required: such as that which triggered deep ice ages (see Chapter 3 for the very brief projection of what was occurring on Earth prior to c. 12,000 years ago).

[300] Miller combined various compounds to show that a proto-Earth could conceivably have produced amino acids = protolife via various naturally-occurring phenomena (electrical storms, etc.).

Sun is assumed to be short wave in nature: visible light and the all-important and controversial UV and E UV). Models barely show persisting cloudiness indices yet they should reproduce temperatures latitudinally. Models as such treat IR with wariness. A climate model should show energy flows versus temperature, which would then show Earth's thermal radiation in proportion to absolute temperature's fourth power. If the thermal flow is in error by 20 %, say, then the temperature is in error by only one fourth. The overall effect is quite small compared to clouds. Is a temperature change of large proportions underway in a hundred years going to be due to a Greenhouse Effect? It might be the case. However, climate models can and do use a concentrated CO_2 rate that is over double that of the current production. Industry is going to grow larger in time, thus it is just as well to heighten what will become of Earth in light of all the polluters now, and those of the future, in such a scale projection. This could be the climate modeler's justification for projections in an ever-industrializing world that must free itself of fossil fuel dependence.

But there are two sides, top and bottom, to clouds that may be involved in the reflection of radiation. That clouds can trap IR beneath them and so warm the surface has been mentioned. Yet clouds of whatever origin in the troposphere also have a major role in reflecting some of the Sun's short wavelength radiation (or, visible light, and UV) back into space. That could be a cooling feedback in Earth's case: hence the gravity of concern around the CERN experiment.[301] Depending on the thickness of the cloud, the reflectivity can be from nil to total. We must remember of course that clouds—unless they cover almost the entire planet, such as on Venus—are constantly shifting, forming, breaking up, adding to warming as well as cooling in a regional, local way that may be the starting point for better research. Convection cooling is a dynamic rule: heat buildup is temporary under such clouds in a realistic dynamic. Is the albedo of clouds the same for IR as for visible light and UV? Possibly, but this is not known: perhaps trapped IR is far worse than visible light. Clouds, then, share a role with the Greenhouse Gases (GHGs) and conspire with the ice and snow fields of the high latitudes in reflectivity. The role of clouds in reflecting the thermal [infrared] radiation back to Earth's surface to create warming must also, however, remain in perspective. Most of all in such feedbacks as positive and negative concerning CO_2, whatever its contribution to cloud cover and local global warming, what is often left out of the discussion is how much CO_2 is reabsorbed by the firmament whatever

---

[301] J. Kirkby et al, "Role of sulphuric acid, ammonia and galactic cosmic rays in atmospheric aerosol nucleation." Nature, 476, 429-433, 2011. Note especially Fig. S2c taken from supplementary online material which showed the increased particle creation rate. Color coding in this figure shows the proof. In an early-morning experimental run at CERN, starting at 03.45, ultraviolet light began making sulphuric acid molecules in the chamber, while a strong electric field cleansed the air of ions. It also tended to remove molecular clusters made in the neutral environment (n) but some of these accumulated at a low rate. As soon as the electric field was switched off at 04.33, natural cosmic rays (gcr) raining down through the roof of the experimental hall in Geneva helped to build clusters at a higher rate. How do we know they were contributing? Because when, at 04.58, CLOUD simulated stronger cosmic rays with a beam of charged pion particles (ch) from the accelerator, the rate of cluster production became faster still.

its source of release. Also left out is that ubiquitous plant life—overwhelming animal life in complex and simple forms by vast margins—uses CO2 as a fuel to make sugars and ultimately produce breathable air for animal life, is undeniable scientific fact. That CO2 actually enhances the robustness of plants as such is also noted. CO2 is released by for instance the production of cement, is also (as the Biosphere II [302] reanalysis showed at Columbia University) readily *reabsorbed* by cement.

The effects of cloud cover on temperature is a familiar experience and this differentiates Earth from Venus in a sharp manner, astronomically. Areas without cloud cover such as deserts or habitual high pressure systems (such as in the US Southwest) see the temperatures drop sharply at night. In areas with cloud cover, such as the Mongolian Low or Greenland Low Pressure systems the drop in temperature is less on average from day to night. Daylight in the former case yields extremely high daytime temperatures and in the latter, more or less evenly-balanced daylight temperature-to-nighttime temperature ratios.

"Clouds . . . could be the most dominant influence in the radiative budget of the lower atmosphere . . . but adequately taking them into account raises many problems," according to John Houghton.[303] John Houghton was, notably, one of the first strong adherents of the IPCC and global warming by CO2 as a potentially potent threat. Whatever one makes of this, his point about clouds as an Earth climate influencer have been taken seriously in all quarters. Recently, as some have taken from observations and experiment [304] clouds at low level (1 to 3 miles high) have a most evident short wave-radiation feedback We have seen how ENSO has been a motor of cloud dispersal along with the ever-eastward Coriolis Effect, and what short-term to quasi-long term climate changes they can wreak. Strong El Nino's from ENSO have been known to effect enormous rain cloud dispersal across otherwise "usual" high pressure systems for whole, expected "sunny and warm" seasons in the United States (east of ENSO). As shown in Chapter 3, that such rain cycles have been tied to solar activity—particularly weak solar activity—is antique information that bears strenuous re-examination.

Clouds as a radiative budget influencer have been cited in detail regarding the act of observation as such. Are they important to climate change or not? First is shown the average, journalistically/ political action committee-labeled climate change skeptic's (or denier's) view:

The purpose of the present note is to inquire whether observations of the earth's radiation imbalance can be used to infer feedbacks and climate sensitivity. Such an approach has, as we will see, some difficulties, but it appears that they can be overcome. This is important since

---

[302]   Ibid Severinghaus, J.P., *et al,* 1994

[303]   Houghton, J., *The Physics of Atmospheres* (CUP: 2002) 3rd ed., p. 41

[304]   Such as the Earth Radiation Budget Experiment (ERBE), and data from CALIPSO lidar (CALIOP) and CloudStat radar (CPR)

most current estimates of climate sensitivity are based on global climate model (GCM) results, and these obviously need observational testing.[305]

And this in turn by the journalistically-labeled/political action committee-labeled climate alarmist (or warmist) view has been critiqued to have more of a correlative rather than causative nature.[306]

As so-called "skeptic" studies of climate sensitivity with clouds in mind (from ENSO etc.) as causation, as in cause and effect, the following was found:

. . . However, warming from a doubling of CO2 would only be about 1°C (based on simple calculations where the radiation altitude and the Planck temperature depend on wavelength in accordance with the attenuation coefficients of well-mixed CO2 molecules; a doubling of any concentration in ppmv produces the same warming because of the logarithmic dependence of CO2's absorption on the amount of CO2) (IPCC, 2007) . . . . This modest warming is much less than current climate models suggest for a doubling of CO2. Models predict warming of from 1.5°C to 5°C and even more for a doubling of CO2." . . . As a result, the climate sensitivity for a doubling of CO2 is estimated to be 0.7 K (with the confidence interval 0.5K-1.3 K at 99% levels). This observational result shows that model sensitivities indicated by the IPCC AR4 are likely greater than than the possibilities estimated from the observations.[307]

This was countered by Kevin Trenberth [308]—a so-called warmist/alarmist—as follows:

Numerous attempts have been made to constrain climate sensitivity with observations. [309] While all of these attempts contain various caveats and sources of uncertainty, some efforts have been shown to contain major errors and are demonstrably incorrect. For example, multiple studies . . . separately addressed weaknesses . . . The work of [Trenberth] . . . for instance, demonstrated a basic lack of robustness in the LC09 [310] method that fundamentally undermined their results. Minor changes in that study's subjective assumptions yielded major changes in its main conclusions. Moreover, Trenberth *et al.* criticized the interpretation of El Niño-Southern Oscillation (ENSO) as an analogue for exploring the forced response of the

[305]   Lindzen, R.S., Choi, Y.S., "On the determination of climate feedbacks from ERBE data," *Geophys.Res. Lett.* **2009**, *36*, L16705, p. 1

[306]   Ibid

[307]   Lindzen, R.S, Choi, Y-S, "On the Observational Determination of Climate Sensitivity and Its Implications" Asia-Pacific *J. Atmos. Sci.*, 47 (4), 377-390 (2011)

[308]   Trenberth K.E., *et al,* "Issues in Establishing Climate Sensitivity in Recent Studies," *Remote Sens.* 2011, 3, 2051-2056

[309]   Ibid Lindzen-Choi (2011) and Spencer, R.W,; Braswell, W.D., "On the diagnosis of radiative feedback in the presence of unknown radiative forcing," *Journal of Geophysical Research,* Vol. 115, (2010)

[310]   Ibid

climate system. In addition, as many cloud variations on monthly time scales result from internal atmospheric variability, such as the Madden-Julian Oscillation, cloud variability is not a deterministic response to surface temperatures. Nevertheless, many of the problems in LC09 [311] have been perpetuated, and Dessler [312] . . . has pointed out similar issues with two more recent such attempts.

Trenberth points out the dangers of seeing cause and effect in correlative matters:

Accordingly, in any analysis, it is essential to perform a careful assessment of (1) uncertainty in any data set or method and (2) causal interpretations in the fields observed; while (3) accounting for the natural variability inherent in any observed record. Several recent instances in which these basic tenets are violated have led to erroneous conclusions and widespread distortion of the science in the mainstream media. For instance, SB11 [313] . . . Moreover, correlation does not mean causation. This is brought out by Dessler [314] . . . who quantifies the magnitude and role of clouds and shows that cloud effects are small even if highly correlated.

Dessler's work, cited by Trenberth, was critiqued by Spenser and Braswell on clouds and Earth's energy budget as a "misdiagnosis."[315]

In the words of Joe D'Aleo, "the IPCC AR4 discussed at length the varied research on the direct solar irradiance variance and the uncertainties related to indirect solar influences through variance through the solar cycles of ultraviolet and solar wind/geomagnetic activity. They admit that ultraviolet radiation by warming through ozone chemistry and geomagnetic activity through the reduction of cosmic rays and through that low clouds could have an effect on climate. IPCC AR4 [316]summarized their views as follows:

Since TAR, new studies have confirmed and advanced the plausibility of indirect effects involving the modification of the stratosphere by solar UV irradiance variations (and possibly by solar-induced variations in the overlying mesosphere and lower thermosphere), with subsequent dynamical and radiative coupling to the troposphere. Whether solar wind fluctuations (Boberg and Lundstedt, 2002) or solar-induced heliospheric modulation of galactic cosmic rays (Marsh and Svensmark, 2000b) also contribute indirect forcings remains ambiguous.

---

[311]   Ibid

[312]   Dessler, A.E., "Cloud variations and the earth's energy budget," *Geophys. Res. Lett.* **2011**

[313]   Ibid Spencer

[314]   Ibid Dessler

[315]   Spencer, R.W., Braswell, W.D., "On the misdiagnosis of surface temperature feedbacks from variations in earth's radiant energy balance." *Remote Sens.* **2011**, *3*, 1603-1613

[316]   2.7.1.3

And so it remains. Specialists can continue the investigation of methods etc., in the various papers and summaries cited here. We contrast these forays into meteorological science as they clash with knowns and unknowns with the report of de Jager, on the satellite data from the last three decades showing the change of TSI "over a typical Schwabe Cycle is about 0.1 percent, corresponding to 0.25 W m2, or, the global mean value at the Earth's surface. This is an estimate quite small compared with the 3.7 W m2 estimated for a doubling of CO2." Hence the fear of CO2 and the negation of clouds as a climate modulator. Yet the CO2 doubling as a causative for climate change, on the other hand, is tackled by the variability of solar radiation according to de Jager and others,[317] strongly wavelength-dependent as it is. Hence it is another model. For "model calculations show that through dynamic coupling, Spectral Solar Irradiance (SSI) changes can cause shifts in the tropospheric circulation systems and, therefore, change the climate." [318] Other models [319] in combining satellite data and models to estimate cloud radiative effects at the Earth surface and in the atmosphere discovered that a combination of satellite observations (and, unsurprisingly enough, models) that the cooling effect of clouds far outweighs the long-wave or "greenhouse" warming effect.

Is a model just a model, or is it based on observations and tests? Are tests and observations more important? Functionalist scientists tend to think so. In some cases, the complexity of the natural subjects involved can do no more than invite more models to be made, despite the observations and tests. No one knows all the parameters involved in the Sun-earth connection, let alone details of the basic operating principles of climate. This whether looked at purely apart from solar factors or as with many of these as possible.

As mentioned earlier, cloud cover overall may play a small role in short and long term climate modulation compated to EUV. That newer space-based measuring methods aimed at the sun, such as the EVE subsystem, are promising to turn the "primitive" [320] science of meteorology on its head due to EUV research in the stratosphere remains to be seen at this juncture in time.

◊

---

[317] Climate scientist Bas van Geel for example: "History tells us that solar forcing of climate change is probably much more important than as communicated by the IPCC. Based on the paleo evidence we know that the climate system is hypersensitive to relatively small changes in solar activity."

[318] Haigh, J. & Blackburn, M., "Solar influences on dynamical coupling between the stratosphere and troposphere." *Space Science Reviews* (2006) 125(1-4), 331-3444

[319] Allan, R, (2011) Meteorological Applications, 18 (3). pp. 324-333

[320] In the words of Richard Lindzen

Solar forcing and earth oscillations on decadal and multidecadal effects from a gradient flow perspective [321]

Presented here is a plausible if relatively unknown aspect of the Sun-earth connection that builds obliquely off topics taken up in Chapters 3 and 4 (for example the analysis of gradients) and the forcing observations made in Chapter 6. (Keep in mind here the bow shocks Earth sustains when a combined CME/solar flare punch is sustained by our planet.) No conclusive statements are made concerning this research and it is suspect like this very book. In the matters of ENSO and the still-anomalous cause(s) of the so-called Little Ice Age (LIA) the following little-documented and discussed research bears note. Could the LIA be a purely hydrological phenomenon? Earth water's gradient flow in respect to solar forcing and Earth's motion is at present not pinned to the floor of academic grant/institutional funding as much as is the question of cloud cover in global climate, for instance, is. Refreshingly, this area invites astronomical causes for climate variation in a geophysical perspective that begs reflection.

Posited is that solar cycles modulate equator-pole Earth oceanic pumping variations yearly (with an effect on tropospheric air currents) and regionally. (Recall the Hadley Cell from Chapter 3 and 4.) It is further speculated upon here that (presumably "normal") solar maxima attenuates (reduces and disturbs) and solar minima amplifies (increases or intensifies) semi-annual terrestrial gradient flow variations. The changing frequency, or the "pulse," of the differential modulation controls the multidecadal oscillation. It is implied here that solar and Earth oscillations are somehow tied.

Multidecadal terrestrial oscillations come about in a cumulative manner from gradient and flow shifts aliased differentially from the solar pulse position by dominating terrestrial cycles (annual, Quasi-Biennial Oscillation [QBO] [322]) such as the north-south asymmetry and ocean-continent heat capacity contrasts.

What are the "mechanisms" involved? They could be gravitational and space-time insolation tides. The surfaces of Earth's shells wrap on themselves, putting constraints on the Moon and the Sun's effects. How the mechanisms function in detail is described by N.S. Siderenkov.[323] Siderenkov speculates in another paper [324] on the close correlation of the decade variations in

---

321   The following is excerpted from the work of Paul Vaughn.

322   QBO is a stratospheric-tropospheric connector to Earth climate. (Recall how the SDO subsystem (EvE) is investigating EUV and UV.) and how CME/solar flares can deliver charged particles in the stratosphere. See for example, Liess, S., et al, On the relationship between QBO and distribution of tropical deep convection, *Journal of Geophysical Research*, Vol. 117, (2012)

323   Sidorenkov, N.S. (2005). Physics of the Earth's rotation instabilities. *Astronomical and Astrophysical Transactions* 24 (5), 425-439 http://images.astronet.ru/pubd/2008/09/28/0001230882/425-439.pdf

324   Sidorenkov, N.S. (2003). Changes in the Antarctic ice sheet mass and the instability of the Earth's rotation over the last 110 years. *International Association of Geodesy* (Symposia 127, 339-346)

Earth rotation with the mass changes in the Antarctic ice sheets. (Again, we recall Kevin Trenberth's words on correlation not equalling causation.) Still the matter bears a look. The redistribution of water masses on the Earth includes changes in the components of the Earth's inertial tensor and causes the motion of the poles and changes of the Earth's rotational speed. Stressed in these considerations is the distribution of water over the planetary surface—something that is remarkable and unique to Earth's makeup compared to all other Solar System planets. We consider what occurs to the two (upper and lower) undersea oceanic "pump" that spread water around the continents (for example those carrying the Gulf Stream from the south toward the north poles).

Nonstationary multidecadal terrestrial variations are synchronized with changes in the spacing of decadal clusters of terrestrially-asymmetric semi-annual equator-pole flow effecting amplitude. The majority of recent multidecadal terrestrial variability is due to natural space-time aliasing differential solar pulse-positions by terrestrial topology over basic terrestrial cycles, including the year. Pulse "signals" sampled in time are indistinguishable from one another over time and space (spatio-temporal) and hence the "noise." [325]

Whatever merit the above considerations have, not only to climate change in normal or natural (non-human) aspects, but also to orbital forcing matters, is welcome to inspection by each and by all.

Sunspot groups: a view over the last eleven solar cycles

Though this book has striven to show that the relationship between whatever the solar magnetic flux is, versus the sunspot number, are not tied as phenomena, sunspot group counts are given here for a set time period. Ignored is whether or not TSI is calculated as variable or invariable, or if the minima-making (poloidal) field is considered stable at the boundary or not, or even if TSI and sunspots bear a nonlinear time relationship (or not) with the magnetic field (toroidal maxima "machinery"). What the images of the cycles are intended to show in this context is how much more magnetically calm the Sun was before approximately 1920 compared to the recent epoch of from c. 1920-2009 as an annex to Chapter 7 of this book.

---

[325]   Le Mouël, J.-L.; Blanter, E.; Shnirman, M.; & Courtillot, V. (2010). "Solar forcing of the semi-annual variation of length-of-day," *Geophysical Research Letters,* 37. Bear in mind that these variations are only tenths of a degree in absolute temperature (Kelvin). Even though they appear to have fluked out somewhat in achieving sufficient focus to recognize the pattern with compoundly suboptimal focus, Le Mouël, Blanter, Shnirman, & Courtillot (2010) have supplied absolutely crucial help with their seminal observation of the missing link that bridges the lower and higher timescales. The primary poor functional numeracy roots of the collectively-compromising comprehension-bottleneck appear to be severely inadequate common knowledge of cycle and phase relation fundamentals and (2) widespread mainstream failure to fundamentally differentiate marginal spatial & temporal distributions from joint spatiotemporal distributions (spatiotemporal version of Simpson's Paradox).

Cycle 12 is given as the starting point even if Cycle 1 is reported as beginning in 1755. The reason to ignore cycles before c. 1860 here is twofold. The first is that data for sunspots and perceived flux is more solid after 1860, especially photographically (c. 1872 onward). The second is that Cycle 12 coincides roughly with the end of the Little Ice Age (c. 1880). Noted also is that Cycle 17 contains the epoch marking the negative end of the Hallstatt Cycle. Hinted in the diagram of Cycle 23 and 24 is a possible return to solar activity more like what was seen in say, Cycle 14 or 15 but that, of course, is pure speculation.

The diagrams below show the starting year for each cycle listed in the graph, the end terminating approximately with the beginning of the new cycle.

Cycle 12 and 13

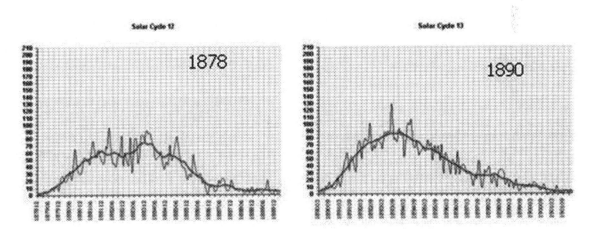

Cycle 14 and 15

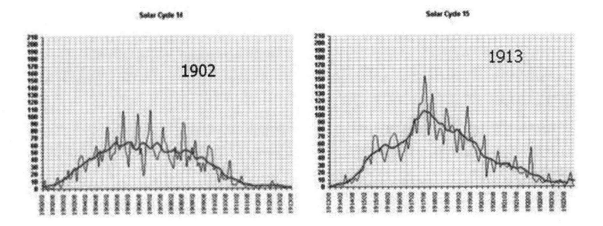

Cycle 16 and 17

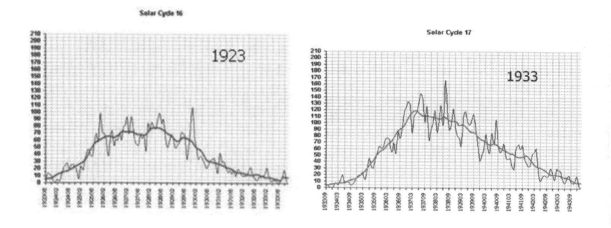

Cycle 18 and 19

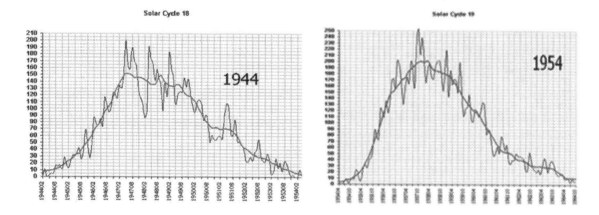

Cycle 20 and 21

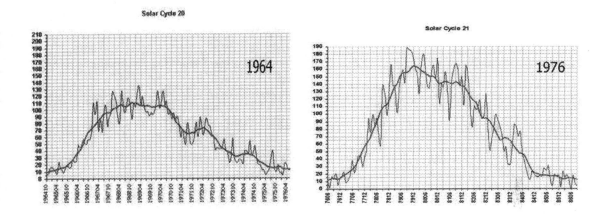

Cycle 22

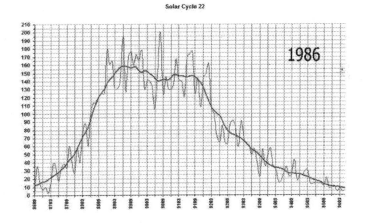

Cycle 23 and 24

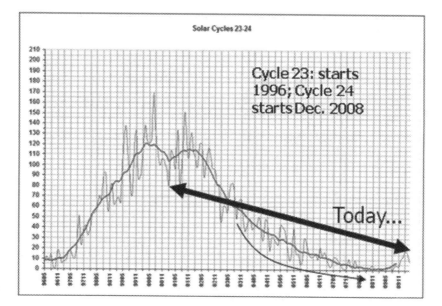

In any general discussion that might ensue from this side by side-placed information, the most noteworthy are two views in particular that can be discerned from a cursory examination. One is that the beginning of the current cycle (Cycle 24) and the ends of Cycles 12 and 13 resemble each other the most in amplitude, overall. As usual, the higher the sunspot count, (usually) the greater is the flux index though there is perhaps no direct relationship between these. The other view is the "beginning" of the (so-called) Modern Maximum, which in de Jager and Duhau begins on the Sun in 1924 (see Chapter 7) at the Gleissberg transition point (between Cycle 16 and 17)—a transition that may have marked a turbulent strike upward for an extended solar maximum. Note the looping, wide amplitudes in the flux in Cycles 18 and especially 19, then Cycle 20, betray a maximum that is prolonged, even as it bleeds out across all other cycles until 24. It stands out in stark contrast to cycles before Cycle 14.

Not pictured are Cycles 5 and 6 which roughly cover the Dalton Minimum period in the 19th Century, and which resemble Cycles 10-13 which were lower overall. Cycles 8 and 9 in the 19th Century were—like some periods in the 20th Century—quite strong. However, they were hemmed in by weaker overall cycles such as Cycles 5 and 6, and Cycles 10 through 13. Being the product of less systematic and photographic study, these are not overly commented upon, howsoever much climate descriptions of those days could be compared to any impending incidents that may occur over the next dozen years or more. In the de Jager context, the Dalton Minimum (c. 1795-1820) is not even considered a deep minimum episode.

That the impact in any long-term context of similarities in weather and so on with the 19th Century could be made must also be thought of in the important context of how much anthropogenic warming would contribute, in such a circumstance, to artificial Earth warming. That is to say, the climate impact of Man-man (say, Northern Hemispheric) warming might stand out in some quantifiable manner, should the solar-geomagnetic input into Earth climate be more readily substracted due to weaker solar indices as witnessed in the early and late 19th Century. Also to ponder, overall, is that most of these weaker or more "normal" or "regular" solar cycles in the 19th Century and earlier were in the last leg of a negative ("cooler, more tending to grand minima") phase of the still recondite Hallstatt Cycle (revisit the musings upon this in Chapters 3 and 7).

Pan-hemispheric weather incidents over the last decade: a look at local effects taken in tandem without statistical smoothing

Albedo reflection off of Earth in the year 2000-2001 timeframe has cast the last decade and one half in the light of potential global climate cooling. This negative feedback of the sun's surface reflectivity off of Earth as discussed earlier alone could account for a major amount of the local global cooling reported from various areas in the last decade and a quarter. Again, since no tools yet exist for local and regional quantification across large swaths of Earth there is only mean averaging to show as proof. No figures for either death by fire or ice will be given as they most likely would be equally invalid.

A cursory inspection of the last half dozen to three years or so shows a great deal of odd circumstantial evidence, as they would say in court, such as a growing lack of solar "surface" brightening or, the lack of large sunspot groups. Look at the hemispheric weather, just to turn away from very low sunspot counts—a non-bearer lately of activity centers on the sun. The year 2009, for instance, regarding summer in the US Northeast, showed pretty much of a non-summer in some respects with uneven cooling into June and heavy rains. These local effects were not strong enough to dampen statistical averaging of punctuated high-heat days that skewed averages into so-called "record warm" descriptions (read the Llanos report below). These record warms, for example, showed one headline proclaiming 2010 as being the "fourth warmest summer in recorded US history" (probably since formal record keeping began in 1875 although this has not been substantiated by me). The summer was dominated by global west-east wind flow on a large scale and then punctuated every now and then by a global north-south windfall. This was somewhat like the summer of 1666. In 2010 as in 1666, severe droughts in some areas, lack of convection cooling, and wildfires were dominate (in 1666 the city of London burned to the ground).

Unlike in the Maunder Minimum, so far as we know, Russia took a hard beating with uneven droughts and fire that year. Killing Russian droughts in the summer of 2010 caused a trade imbalance on wheat and other grains, where China, the usual customer of the Russian crop, turned to the US for importation, causing a rise in US prices due to scarcity. This was due to a late summer harvest in the US that was much less than anticipated (corn, wheat) than the usual and recent bumper crops of the 1980s and 1990s. The US aimed at a global supply for a growing international market. But China's demand for grain alone meant that the US reserve supply for 2011 actually approached a billion tons which was critically low. It probably spiked up the cost of dairy products in the US as a side effect. Most people who use dairy products are familiar with this by now. China experienced bad coastal weather reminiscent of the Maunder Minimum period that year, as well.

"Dustbowl drought" began in Kansas, Oklahoma and Texas across late summer 2010 into spring, summer, and fall of 2011. Oklahoma was drier in the four months following Thanksgiving (2010) than it has been in any similar period since 1921. That says a lot in the state known for the famous 1930s Dust Bowl, when drought, destructive farming practices and high winds generated severe dust storms that stripped the land of its topsoil. This was in a time, we should note, before the Modern (solar) Maximum began and was ensconced in a rather cool period culminating around 1945. Neighboring states were in similar shape as the drought stretched from the Louisiana Gulf coast to Colorado, and conditions worsened, according to the US Drought Monitor. The area in Texas covered by an extreme drought had tripled in a single month to 40 percent, and in Oklahoma it nearly doubled in one week to 16 percent, according to the monitor's March 29 update. (Only Northeast Oklahoma was free of drought due to heavy snows.) Mr. Michael Spradling, the president of the Oklahoma Farm Bureau, said "many wheat farmers considered just plowing under their fields and switching to another crop." Drought relief came to parts of the US south by September. There was suddenly 14-20 inches of rain in the Gulf of Mexico region on September 3 to 4, 2011. But parts of the worst-hit areas, for instance in Texas, never got the relief that stretched into 2012. While the south and west burned in 2011 the north drowned. There was record flooding in the so-called rust-belt to the northern New England border. (Lake Champlain exceeded pre-National Weather Service reports of lake rise and flooding exceeding 1869 measurements). Other than burning the US south principally in the spring of 2011, the year and many unhappy southerners witnessed at least eight major tornados between March-June, culminating in tornado warnings and actual incidents of them from up to the Maine border with Canada in the late summer (Hurricane Irene for example). As we might recall the Dust Bowl period was also characterised by terrific cyclonic activity. The death toll from spring storms in US that were hurricane and tornado-related in 2011 exceeded that from 1953 (+ 300 in 2011 . . . 1953 had a total of 519 deaths). This coincides with low sunspot / electromagnetic activity on the sun in 1953-54 timeframe and which was also a period of high cyclonic activity in the US east for sure. Hill and mountainsides vanishing into river basins due to oversaturation, forcing a change of state from solid soil and mineral into fluids, was reminiscent of the far west. Such mixing up of debris in this deluge, which with its consequent raising of the water table, presently runs risks of introducing bacteria and fungus into sweet water sources. Emergency and water testing services are gaining wide popularity.

There were unexpected and violent, destructive hurricanes in the north not seen since the 1930s or even earlier. I make a brief correlation here between this and the drought indices in the Texas-Oklahoma region at approximately the same time. In terms of pan-hemispheric considerations, I point out the Japan tsunami of April 2011 (coinciding with the near-epic earthquake of c. R-9 on the Richter Scale). Over 15,000 people perished, though this is not connected with Hadley Cell theory or hurricane-like storms. It chimes in with the December, 2004 "Christmas" tsunami in the Indonesian region of the "ring of fire" where 250,000 plus persons died, and if we recall orbital forcing factors, the insights of Rhodes Fairbridge and others on tectonic plates in this potential regard, we see actual disasters that may be working in a linkage of some sort. But here I overly speculate.

Storm magnitudes as force vectors measures (Fukuda, Saffir Simpson Scales) approach the 1920s levels in the US. The National Oceanic and Atmospheric Administration (NOAA) Physical Sciences Division in their El Niño/La Niña page discussed the impact of ocean temperatures on tornado events that year. [326] (They conducted) "a preliminary investigation of the co-variability of 20 severe spring (March-May) tornado outbreaks over the contiguous US and phases of the El Niño/Southern Oscillation (ENSO) during the past 100 years. (Their) assessment reveals that the relation of ENSO with major destructive tornado events appears to vary with time." Compare this note against the section on "Solar forcing and earth oscillations on decadal and multidecadal effects from a gradient flow perspective" above.

Meanwhile, strong monsoons in Pakistan drowned out the crop region there in 2011 while Russia burned to a crisp. Too much water south: not enough, north. Here is a case of uneven rains and a scenario like that described in the Hadley Cell circulation. The mainstream press mainly reported on the "devastation of fire" in areas of the globe that were steppe-like or near desert that summer and autumn. Downplayed was the role of lessened crop yield and destructive flooding. Killing droughts got relief at the wrong time. There were even destructive monsoons in Nigeria late that year, probably from the same Kamikaze winds that had wafted in a most unfriendly manner across the Indian Ocean into Pakistan.

The Southern Hemisphere showed a similar skew along the lines of the Hadley Cell circulation description offered by van Geel et al in Chapter 3 in the year 2010. South America, which was not on radar screen in Maunder Minimum so little comparison can be made to it with now, bears examination. Presses and news stations in the US seemed to have all but ignored the goings on in some of the worst summer weather in the Andes and in Chile and surrounding nations in decades or even centuries. Killing frosts in the South American 2010 winter (Peru, Argentina, Chile) were the hardest reported in some cases, ever. Rural populations in mountainous areas suffered greatly and relief was hardly forthcoming to extremely poor peoples who were unaccustomed to such cold snaps. Frozen rivers, frosts, and snows were reported in areas that had hitherto "never seen it before," to repeat a sad but by now familiar refrain. Coffee prices

---

[326] Detailed analyses of surface and subsurface temperature conditions related to such events can be looked at in the work of Roy D. Spenser.

went up, unsurprisingly enough.[327] In New Zealand spring frosts and snows killed millions of sheep in the 2010 winter there. Again, these reports never made it to NBC, CBS, or ABC even if the BBC regularly reported on them.

Winter snowfall in the US came early and deep in 2010. Early ski resort snowfalls for northern Canada were reported in September and probably matched if not exceeded 2009. Early ski resort snowfalls for Switzerland also came in September in the Alps. So the early snowfall that year with a good amount into April was pan-hemispheric. Winter snowfall in the US actually broke records in states across the winter 2010-11. Snowfall build-up was high and ice melt low into February 2011 and early March. Record cold (below zero) was reported in many states across winter 2010-11. There even was some snowfall in Los Angeles, California in February 2011.

For the second year in a row, the winter temperatures when averaged across the contiguous United States came in below the average temperature for the 20th century. This marks the first time since the winters of 1992-93 and 1993-94 that two winters in a row had been below the long-term normal, and it made for the coldest back-to-back winter combination for at least the past 25 years.

The change from 2011 to 2012 from winter to spring showed lower US continental snowfalls but Alaska (Nome) got "iced in" and required the aid of Russian oil tankers to deliver life-saving fossil fuel. A 24-foot snowfall in the city near Fairbanks on the lower limb of Alaska revealed snows not stopping for weeks. Snowfall also occurred higher up in the US west coast than "normal," where events like five inches falling in Seattle had not been seen "in generations."

This snowfall and cold followed a pan-hemispheric trend. Europe had enjoyed a relatively mild winter until late January when a Siberian cold front came in from the Northeast, plunging temperatures to -25 degrees Fahrenheit and even lower in some places. Eastern and central Europe were hardest hit. Widespread snow in northern and central Italy disrupted travel. Trucks were banned from roads in several regions, including Tuscany. Air travel was cancelled. Mountain roads on the French island of Corsica were shut by heavy snow and 50,000 customers lost power. Europe's Siberian freeze spread to France and Italy, with snow covering Corsica, Bologna and Milan, while the European death toll topped 100, Snows in the Balkans had people tunnelling 15 feet into it. Food was sent in as emergency support.

By February 7, 2012, over 350 died in Europe in cold related incidents. Freezing occurred in London for the first time in years. Snow fell in Rome, the first snow in 26 years. The BBC estimated 30 people in eastern and central Europe froze to death in Bulgaria, the Czech Republic, Latvia, Poland, Romania, Slovakia and Ukraine in snows and cold not seen in generations. Temperatures were so low in parts of Romania that parts of the Black Sea near the shoreline froze. Istanbul,

---

[327] Maxwell House increased it by 25% in 2011-2012.

Turkey, which rarely gets snow, got 20 inches and double-digit sub-zero temperatures in February. Temperatures in northern Greece hovered above freezing. Africa was not spared. Snows in Libya happened for the first time in such a heavy manner since February, 1956. To their credits, NBC, CBS and ABC made passing references to these trends, unlike in 2010-11.

Winter came early in 2012 in the Southern Hemisphere. On August 7, 2012 snow fell in South Africa for first time since 1968. They have had such snow only for 22 days over the last 103 years. But by October, 2012 in the US, a part of the Northern Hemisphere, after a long, dry, drought-filled summer across the entire nation (south, north, and west) winter suddenly appeared. It left in its tracks a sad report from pork farmers that it was not worth the cost to raise pork, seeing the glum expected returns and prohibitive costs on grain. Prices will certainly rise for pork at least. Talk of making fuel from corn is turning into action to prevent it from being taken from human mouths. The summer of 2012, probably doing to the bread basket of the United States what the summer had done to Russia the year before, had journalists doing the take on statistical averaging in the continental [328] US as having had the "warmest year on record":

'Warmest year' looking more likely for 2012 across continental US [329]

With less than three months left this year, it's looking increasingly likely that 2012 will go down as the warmest year on record in the continental United States. Moreover, six of eight scenarios charted by the (NOAA) center have 2012 ending warmer than any other year in records that go back to 1895. The only scenarios where that would not happen are if the last quarter is among the 10 coldest on record . . . This year has already seen the warmest March and July on record, and, except for September, every other month was in the top 20 warmest, weather. com noted.

Again, we look at statistical averaging to see how misleading such news items can be. "This year has already seen the warmest March and July on record" can easily be taken apart like this: March for many areas of the continental US had one week of extremely abnormally hot weather around the 90 degree Fahrenheit mark. That is, c. seven days, experienced across the nation and in other nations.[330] Otherwise the month had been cold, dry, and often sunless. As such, a single week of freak weather, whether experienced at the top of the month, the middle, or at the end of it, skewed the monthly average out of proportion to what most of the weather was like, for most people, for most of the month. Nowhere in the record-breaking literature at the to p level of a record spring heat is it identified or not that the warmth was uniform or nearly so

---

[328] This conveniently excises Alaska from the picture, which perhaps had snow on the ground all summer in places.

[329] Miguel Llanos, NBC News

[330] State of the Climate Global Analysis March 2012, National Oceanic and Atmospheric Administration National Climatic Data Center

for 30 days. In my personal experience in Vermont it was one week, the temperature extremely high. April saw a rapid return to cold, dry weather. In other words, I hazard most of March, 2012 was actually cold in *local*, or *regional zones,* harsh realities that statistical niceties in official government agency reports will never reflect. Persons such as myself saw and sweltered while trees and plants remained free of buds, these latter not fooled by the freakish burst of heat. The entire measurement is reminiscent of how a few extremely high test scorers on SAT tests can warp a whole sex into seeming like a superior entity.[331] The reality is obscured withal in either case unless for you, the reality is statistical averaging. The report from NBC showed some balance, leaving at the end of it the proviso that this miracle of statist mean averaging could be upset if the final few months of 2012 showed a significant cooling. But even this *cooling* could be just for a single seven days spread out unevenly across 50 states in one month, skewing the temperature average cooler (not the actual conditions) for a whole month when actually, most local, regional areas might experience mild, sunny weather. If it is any consolation, Δ (Delta) C14 measurements—as this book has shown—are also poor reflections of solar operating reality on climate in spite of how much this book appeared to tout its merits.

Of perhaps worthier note for the end of the year 2012 might be an examination of winter and hurricane season (July-December) storm intensities and lengths of duration as a sign of degraded weather in a climate system undergoing a rude and hard-to-fathom change. For instance, storm intensities in the autumn of 2012 on the Eastern US seaboard revealed coastal-impact patterns not seen for hundreds of years, back when the norm was powerful cyclone activity that could last for whole weeks, with attendant destruction. That storms of such magnitude reach higher northward from usual southern patterns has been touched on elsewhere in periods of reduced solar magnetic activity.[332] Coastlines in Scotland were pushed up and back and coastlines in the Mediterranean rearranged in North Africa in the Maunder Minimum. The energy transport in ocean-atmospheric matters could shift in low solar periods, causing orographical-topological features to take on new significance as added forcing vectors (this however is beyond this book's reach). Measured low pressure for the storm Sandy recorded by the US weather service exceeded anything recorded so far in the US republic (recall the discussion on altered gradients in Chapter 4). Coastal rearrangement was reported. The last time such a storm with such low pressure approached the US coast was perhaps when Blaise Pascal, for whom the unit of practical pressure is named, was still alive—during the Maunder Minimum. Water levels in the city of New York at Sandy's peak exceeded levels seen there

---

[331] See Sommers, C. H., *The War Against Boys, Atlantic,* May, 2000. Re-analysis of statistical methods versus results in one case showed that boys' overall higher test scores than girls' on SATs were due in the male case to a small handful of top male test-takers (whether honest or cheaters) skewing "the peak" higher in overall proportion to girls. Actually, most boys scored worse than most girls. Hence, this prominent feminist's article title. Selected or self-selected boys compete harder on the majority of boys (and of course girls) in the competitive drive attributed perhaps more to male glands than to any innate superiority. Female testers tended to a mean, overall "middle" average.

[332] Ibid, Soon -Yaskell (2004)

from 1820: significantly, toward the end of the Dalton Minimum, a kind of non-grand low solar phase we might soon find ourselves in (if we have not already been there for awhile). That a Dalton-like episode could be on the ascendant would be consistent with the Hallstatt Cycle's behavior in its positive phase (see Chapters 4 and 7). Note that the destruction Hurricane Sandy caused was perhaps the most costly in US history. Generally, uneven spread of warmth versus cold areas locally and regionally, dragging with it rains for weeks and snows of great depth for months, did not fit into convenient patterns then and nor do they, now. For example, snows etc. occurring southerly of areas contra areas in the high north experiencing abnormal warm have been ongoing for some time as if a sort of polarity were spinning or reversing. If solar activity at solar maximum by the mid-to end of 2013 is weaker than usual, it would only be to live through autumn-to-winter storms of equal intensities and durations in the years following this to get a picture of a cooling earth. Ironically, we hope in that case that the CO2, water vapor, and other GHGs budgets remain in reservoirs for longer than usual to keep us warmer, locally and regionally, overall. Hardly would the wish be for "global warming" to then vanish.

Meditations on any coming cold

At this book's beginning I made much of the hopeful impending period when newer thinkers and heroes would push the envelope of space research and exploration. We can only hope such intellectual and perhaps just pure adventurers take up the challenge. But the most of us are hardly so gifted or bold or both. This ends my pean to the very young and the as-yet unborn. What does the near future have in store? Some of the Bible's teachings tell us that the house of the future is not the dwelling of those presently alive. This means all that much more care we must give to children and to the future home they will one day inhabit.

It is implied here that global cooling is a potentially worse-case scenario for the ever changing biomes and microclimates we live in on Earth, populated as they are. I am quite aware of the long and serious debate that what is causing uneven weather and its attendant anomalies is due to GHGs and not to cooling at all, from any quarter, anywhere. This is true especially in light of droughts and flooding. But the actual truth of such things is perhaps beyond us. Unless we are not absolutely certain of the one, so can we not dismiss the other. Keeping an open mind is paramount. Excessive global warming is not harmless, as in areas of Earth increased warming poses its own challenges. For instance, that climate change can in some parts of the world be connected to increased warfare indices has been recently and quantitatively shown.[333] Carbon Dioxide as well as monoxide in concentration warps and could kill in heavily industrialised areas. A pressing need for cheaper air conditioning in hot climates where most of the world's poor exist instead of live would contribute greatly to human well being there, as such heat contributes greatly to diminished health and activity. And there are the deadly regional droughts under scorching conditions. That the other dynamic, increased global cooling, can bring about widescale catastrophe demands methodical and quantifiable—not just historical

---

[333] Ibid Hsiang *et al* (2011)

record—proof. Of this I am painfully aware. In any case, if a mechanism for even preliminarily identifying times of longer cooler (less active) solar cycles vs. longer warming (more active) solar cycles can be achieved, so much the better. This has been both the main focal and pivot point of this book.

In one sense the Sun *is* closer to Earth than ever before and especially since the beginning of the satellite age—coincidentally the same month I was baptized in Lynn, Massachusetts, USA: October, 1957—the three customary months after a July birth in one of the strongest solar flare-active Julys ever measured by modern living humans. The blistering weather and hemispheric drought that year coincided with the science-drenched International Geophysical Year (IGY)—to make many understand that the Sun affects us with its power in a direct, visible way. We should keep up the science momentum of that year.

On or near the day I was baptised the world population stood at 2.888 billion souls, up since then to seven billion by November, 2011. This is nearly four billion more humans in a tad more than 50 years. Hence my early concern with painting a picture of "deep time." We in the present cannot fully comprehend the deep past as well as the future. Think on this when we compare such a huge global population jump clashing with any rough climate events that may or may not have marred the face of Earth since the population was less than a billion, or what might come again. And then, this population was much more widely spread out. The world since "time immemorial" suddenly broke two billions in 1940. My mother, still living when this book was begun, had already been married for three years. The principle contributing expert to this book, Cornelis de Jager, was a 20-year-old studying physics forced to hide in a room in an observatory in the Netherlands from the German army due to their desire for his technical and scientific training. It was around the time we discovered that we could do some of the things that Sol could do—if in miniature; that is, split atoms. As such, it also gave us an intriguing closeup glimpse of what kind of terrible power a star can wield.

Before then the world population was flat for a long time indeed, only breaking one billion around 1800. Farther back the numbers get very fuzzy. But we are confident that at no time before AD 1800 did the human world population exist greater numerically than it does now nor in such concentrations in parts.

We have been taught in many large and successful parts of the West that progress knows no bounds and have come to accept it, sometimes in an exceptionally exquisite passive sense. That is, it is accepted that progress will not stop: a dangerous assumption. We sit, our very big human population since circa 1940 on a small heavenly body that geographically has been quite well mapped, generally well explored, and universally populated by us. We certainly do not know all the details about our little house yet, but that is another matter. Earth climate is still a mystery in many ways. That we will go *around* (if not on *past*) our star, Sol is somehow tacitly understood, if not openly expected, so soon. But the chance to be able perhaps to do so has been sooner and not later.

◊◊

Speaking practically, we are bound to this earth in the near term. In the very real world of tiny separate contacts and values, the world is still one vast and far in between where Nature can easily separate and damage. Widening any destructive path, one way or the other, it can quickly or slowly destroy. Any natural return to a cooler Earth reminiscent of a time when the world was numerically smaller in terms of people invites this thought. Thinking smaller and smarter could become a necessity and not a fad.

The predictions of de Jager and Duhau leave two avenues of approach depending on the circumstances outlined earlier throughout this book. One is a return to "regular" or "normal" solar activity as was witnessed from the mid-18th Century to the early 20th. The other is a return to a deep minimum phase, though this has been downplayed in the recent research literature by de Jager himself due to a re-examination of the Hallstatt Cycle involved. But if this latter proves false, in a possible coming of an extended deep cold period on Earth where the reduced-activity Sun plays a large role, we could be challenged greatly to survive as individuals and in tighter communities, be more limited in movement, perhaps very much more limited by choices and varieties of foodstuffs and what we now consider necessities (which are always someone else's luxuries). Even a return to "regular" conditions reminiscent of the 1800s might challenge a world population of several billions. That total human existence is threatened with outright destruction during either time is nugatory. But tempers at all societal levels will gradually be shortened if something like this comes to pass. That romantic fear, "the population bomb" will have a rough "solution." The sick joke is, of course, that when the 2004 tsunami/earthquake struck the Indian Ocean that Christmas a quarter of a million people vanished if not more. A few sound bytes later, a couple months down the road, this was hardly noticed. A few years later it was a boring statistic. As unrelated to global cooling as it is, that is still 250,000 people gone, a significant but vocal and descriptive fraction being northern European tourists who survived where close compatriots vanished. In fine, we may need not necessarily fear human genocide so much when a natural variant may be closer at hand. But any "genocidal" tricks Nature could pull, it being a force and not a *being*, could in turn force similar levers placed in human hands.

Extended, persistent, uneven cold and violent storm tracks across just the northern hemisphere (let us leave off the southern) where rich food is needed to power sophisticated, active, often aggressively intelligent minds, could have a very un-dramatic culling effect, should any artificial human Earth warming by industrial means or otherwise prove to be a null set. Where tempers do not flare and egos not riving asunder there could be a steady return to, I am not glad to say, a roughening of hearts and a grim limiting of expectations. This of course was before the time (a kind of pinnacle, really) of "miracle drugs" and hence an apex of medical and veterinary, if not of all, science in the post-World War II boom of such things. This pre-apex "time" may be on the way back like a car sliding backward down an icy slope. We need preparation and not a flat acceptance, as some would like us to have it, of an attitude of the "survival of the fittest." For a loud and wide acceptance of this under such circumstances would surely come to pass.

Nature is always a potential enemy with no feeling; a "numb' one: the enemy within ourselves is always the one we might be able handle in face of the other. Attitude adjustment carried off well will encourage cooperation and reaffirm the great miracles in science of the past, constant attempts to bolster this and maintain it under the strains of various uneven, hindering limitations. I shall be mundane for a moment to the point of banality. Eleven days without power in a New Jersey suburb after Hurricane Sandy lead to mass protests against utilities that hardly could start supplying a people so used to instant and redundant energy. Violent incidents were recorded on November 12 there. What can trigger anger? In this case eleven days without power in the autumn where Sandy was followed up by a so-called northeaster: a large storm bringing in snow and high cold wind. Or some may not have the coffee they like, or the opportunities to shop till they drop, and not the cuts of meats many take so much for granted in some parts of our wonderful world, and I gravitate towards my own US Northeast again. My 1930s depression-working father's warnings to me as a boy whisper back: "someday only millionaires will be able to afford steak." He said this at a time when steak was getting more economically accessible to many there in the 1970s. I hardly think, with modern agricultural techniques, that things like steak, potatoes and corn will disappear. But they could become more rare in a more populated world with a shrinking or more narrow food supply at hand. Mark well the dissatisfied customer when they cannot get a certain blend of coffee at just such and such a strength, or satisfaction from a meal lacking certain ingredients, such as certain vegetables. Note the anger in New Jersey whilst utilites do the best they can to patch up mangled power grids. The wealthiest will command all such resources first. And in times of want, the happy condition of what wealthy constitutes becomes a very relative thing indeed.

Seasonal expectations of certain fruits, vegetables and in less quantity and quality, would start to become familiar occurrences not witnessed for over a hundred years in some parts, along with other things. Uneven droughts in Russia in 2010-11 and in the US in 2011-2012 will raise the price of pork much higher in 2013 onward whether you want to look at these droughts as death by fire or by ice or not. At least two generations if not more in better-to-highly populated cities of the US Northeast have no direct experience with this, for example. For my 1911-born father root cellars, mason-jarred vegetables and meatless months were a winter routine in southern Massachusetts in the terribly cold 1920s. To leap from him to Albert Einstein (a vasty jump indeed) this latter personage purported that "If the bee disappeared off the surface of the globe then man would only have four years of life left. No more bees, no more pollination, no more plants, no more animals, no more man." They will not of course disappear totally in any climate change for the cooler (though I would be insane to question his mathematics should they do so). Large hemispheric collapses in world bee populations, termed in some instances Colony Collapse Disorder (or CCD) are the result of several factors. Insecticides, crop engineering, shrinking habitats and parasites have impacted the overall health and immune system of bees. The other factor contributing to bee decline relates most probably to side-effects of technology and, it has been noted, solar activity. Remember how, in his isolating C14 as a counting gas, Hessel de Vries found out how sensitive bees were to IR. Bees help us to thrive on our little world. How very sad to be so dependent on so tiny a creature for so much.

In less populated areas around wealth islands like Boston, New York City, New Delhi, India, Paris, France etc. such occurrences are anomalies that are avoidable even if there are among them (like in my rural Vermont) people who choose a seasonal approach to life as a kind of new age mantra that strives to copy the best of the, presumably old, age. This attitude (very relaxing, really) will be a useful one. The operating end of it will mean such an attitude or way of life will have to return to the more populated areas in the US Northeast, growing out of, as it were, this now-unfamiliar and in many cases, unwanted, rural mind-frame. It has to be done peacefully and that will be a challenge. What that says for the agribusiness and restaurant trades (blown all out of proportion since c. 1980 alone) is another matter that, of course, will force a vice-like stress between the profiteer and their well-accustomed customer. The supply of the one will diminish and the "consumer" [334] of the other will drop off. There will be a condition of need that will naturally suppress this. Returns to small, local farms and their products has recently made a resurgence, along with the attitude in not any actual benefit of so-called "green" thinking. Hopefully, along with such efforts, attitudes will change peacefully and without rancor in this transition. Division, testiness, outright anger and even need will never be eight light minutes away: it will be in the next room, down the street, in the hospital, in the office, or perhaps right in your stomach and will always be exacerbated by an eager media wittingly or unwittingly pitting the one against the other as individuals and groups. Religious tabernaculai of all types, law offices; universities and institutes, and government bureaus the elites know (and this knowledge must be transferred peacefully and rationally to non elite businessmen and to the less educated) are actors that often speak neither for God, the law, the sciences, nor necessarily for any of our best worldly interests. The media, police, and militaries are just button presses away from all of these, not to mention the increasing access and din added by personal instant media messaging devices: that is, your hands-on device.

Talk of (potential) vice-like stresses between people. Hurricane Sandy could be a dress rehersal for things to come.

◊◊◊

So-called alternative fuels or energy source exploitation is on the rise. Fossil fuels will be replaced seamlessly by alternate energy resources and it is just as well that we know and expect this. When fossil fuels replaced trees as a carbon source for fuel by the 1920s the trees everybody believed were vanishing since the 1850s then came back. Arguably there are more trees now in the US Northeast, to take a forested part of the developed world, than when the American founders signed the Declaration of Independence. Nothing froze to death in the

---

[334] This term, never a favorite of mine describing myself in the context of markets, may vanish.

seamless transition from living carbon (tree fuel) to dead carbon (fossil) fuel, howsoever much it lead to more societal complexity. Fossil fuels are expensive, have been contentious as to use [335] and profitability for over one hundred years; are not clean, are getting harder to mine and refine, and ultimately are limited in amount. Nuclear power is more common and cheap, but the plants are sitting nuclear bombs.[336] They are always potentially extremely dangerous, given that their wastes alone have to be bound in radiation-proof containers and buried or otherwise concealed for thousands if not millions of years (recall the discussion on radioactive decay times in Chapters 1 and 2: talk of launching this kind of waste into deep space is not idle). It goes almost without saying what happens when nuclear reactor cores cannot be cooled down properly due to natural disasters, polluting air, soil and water near and far, especially as coastlines might be getting more fragile. One of those side "miracles" of being able to crush atoms, nuclear fission is not so miraculous in terms of what we do with the leftovers, which no one likes to think of till radioactive contamination begins leaking back into Nature and thus into our food and water, which it is surely doing more of all the time. Common and heavily relied on now in some nations, [337] it is going to have to "go" altogether.

Wind is seasonal and is, as in the Book of Job, to effect, "capricious," generally powering wind plants better in the winter than in the summer when hemispheric proclivities dictate the strengths of winds off the oceans to the land, for example. It is vice versa in the summer so ultimately, there goes the wind power, summers. Electrical-capacity storage is key. Water power is good if there are heavy water sources available and increased rains in global cooling times hemispherically should keep electrical generating plants well supplied, if not overly so, since even normally hydroelectrically-active areas (say, Quebec, Canada) may suffer uneven desiccation in prolonged solar minima times. Drought is not only a factor of excessive heat but also of unevenly-dispersed precipitation in cold periods. Not every place is so blessed with water in hotter/cooler, sometimes drier, times of "odd" weather patterns. Water may be desperately needed in seasonally drought-stricken or desiccated farm belts in drier areas during extended cold periods, oddly enough, in the times of expected seasonal plant growth.[338] Getting water to such areas may be more of a priority, and this requires precious fuel, probably fossil fuel-powered ocean-going vessels.

---

[335] The American oil embargo of Japan in the late 1930s was a pretext for beginning war with the US.

[336] The recent experience with Japan, where Nature performed a "one-two" combination punch of a near-epic earthquake and a resultant tsunami, ought to make most of us pause to reconsider the fission option.

[337] Such as Japan, which had 54 until recently. Since their island nation straddles one of the most volatile zones in the Pacific's "ring of fire," one wonders why this nation (other than for the energy independence) chooses fission.

[338] One of the first forays were the destructive monsoons of 2010 in Pakistan and later, India, where the dry season was capped by floods stranding millions. In deep solar minima periods such monsoons are only expected to occur more frequently.

Ironically, the Sun—and this sounds counterintuitive—will be a major energy resource and probably the most reliable long-term one during any possible prolonged global cooling (photovoltaic: not thermal—thermal requires far too much water.) The Sun's natural diminution in "magnetically calm times" should have [339] no negative effects on interpersonal thermal home heating devices and advanced photovoltaic capabilities. Solar energy research and development is fortunately being driven in very technically/scientifically-literate Germany, and *in* that country, too, which has massive cloud cover for most of the winter if not year round and generally is a cool microclimate overall. It works there. Thus it will work in similar places (like in the US Northeast) and in a dozen years should be a major part of electrical plant generation everywhere. Niche markets are even leasing panels to homeowners for trying out this method of home heating, and infrastructure in the near future for solar power might become commodity retail items much as mobile phones have become. That is, the infrastruture or equipment or both is less expensive than the storage areas and service supplying them. Profit making off of this must be sooner not later, and high, and I for one am infinitely confident that money makers will find ways to hold us all over barrels for it to create the vital splinter markets off of everything from municipal lighting to home heating, to automobiles, though it is hard to see fossil fuels vanishing altogether for large, worldwide (aircraft, large transport ocean vessel) or near space transport and exploration.

Whatever the outcome of any near-term global effects from a reduced-in-activity Sun, should this ever be the major case, vibrant, powerful, impartial law and the scientific method stand out first in an arsenal of countermeasures to oppose fear and the attendant destruction that usually accompanies it. We must also hope that the effects will not be as catastrophic and Earth-changing in human terms as other books and even this one have somewhat outlined and suggested. As Cornelis de Jager has stressed to me, the phenomenon of a weakened Sun, variable as we now well know it to be, is an entirely natural phenomenon. It is one that humans have probably adapted to so long ago that its overall effect on the human race is easily withstood, should it be influencing climate in ways we do not yet know.

But the last time Earth's Northern Hemisphere was in a "regular" period of solar activity—that is, relatively weak and so, Earth being somewhat cooler—Earth's human population was a few billions less.

◆

---

[339] I say this conditionally. Technology in photovoltaics should be able to exploit at nano levels semiconductors made even out of carbon, to replace the common silicon, with very high efficiency.

Index

A

Á (alpha) waves, radiation, 5, 123 (see also electromagnetic spectrum)

aa index, 134-135 (see also Standard and Lockwood data)

aamin, 129, 131, 135-139, 143, 148 (see also poloid/poloidal)

Abdussamatov, H.I. 144

Abrupt changes in long term solar cycles (see chaotic transitions)

Absolute temperature xxii, 3, 158 (see also Kelvin (s) [K])

Absorptive mechanisms (in magnetic fields in solar and planetary magnetic fields) 61

AC (alternating current) 2 (see also Tesla, N.)

Accelerator Mass Spectrometry (AMS) 23

Active regions (see EARs, centers of activity)

Advanced Composition Explorer (ACE) xxviii

Adams, J.Q., 22

Adiabatic, 85

Aeronautical, aeronomy, 65

Aerosols, 73, 104-105

Afghanistan, 56

Al tusi, xvi

Albedo, Albedo Effect, 45, 65-66, 104, 157-158, 169

Alfvén (Alfven) waves, 119-121

AM (amplitude modulation) 129, 132

Amplitude (in solar activity context) 11, 29, 33, 60, 125, 130-132, 135, 139-145, 164, 168

Anasazi, 50, 52

Angstroms (Ánstroms [Å]) 87

Andes, 42, 44, 171

Annual Conflict Risk (ACR) 56 (see also Hsiang, S.)

Antarctic, 164

Anthropogenic, 169

AR4 (see also IPCC) 160

Archaeology, archaeological evidence, 43

Archimedean spiral, 85, 96, 98, 119

Archives (see ice cores, trees, etc.)

Argentina, 171

Aspects (from astrology to modern astronomy) 106

Astronomical Unit (A.U.) xiii, 61, 107

Atoms (fission) 180

Aurora Australis, 69, 97

- Borealis, 19, 96, 70, 73, 97, 153

Australia, 56

Avogadro, A., 4

Aw-dynamos, 60

B

B (beta) waves, radiation, 5, 123 (see also electromagnetic spectrum)

Babcock, H.W., xxv-xxvi

Babcock-Leighton, xxv-xxvi, 117-118

Background component (of measured TSI) 108

Bacon, F., xxviii

Balmer, J.J., 5, 74

Barentsen, G.W., 24

Bars (see gradients)

Barycenter shift (see inertial motion)

Base length (or basal) lengths of amplitude waves, 11, 132, 138

Base functions of compact support, 1125

Bearers (see sunspots, centers of activity)

Bees, 24, 178-179

Beer, J., 38, 41, 46, 103, 110-111, 113, 116

Bering, 10

Beryllium (isotope Beryllium 1o [BE10]) xxvi, 81, 110-111, 113, 128

Bi-decadal oscillations, 140, 142

Bible, 8, 175

Big Finger Cycles (BFGs) 58-59

Birds (olfactory glands sensitive to Earth's electromagnetic field) 49

Biosphere I and II, xxvii, 78, 159

Black Plague, 27

Bohr, N., xxi, 5, 7

Bond, G. (Bond Events) 32-33, 35-36, 45, 47, 110

Boson (see Higgs Boson)

Boundary conditions, 66-67 (interplanetary shock waves) 96

Bow shock, 61-63 (with plasma) 72, 75, 96, 98, 118, 153-154, 163

Box Turtle, Eastern (as possible Holocene Maximum northward migrant) 36

Boyle, R., 8-9

Braswell, W.D., 160-161

Brewerton (archaeological period) 48

Brightening component (of measured TSI) 108

Britain, 22, 23, 34

British Astronomical Association (BAA) 14

British Empire, 13

Broecker, W., 78

Broglie, Duc de, 1123

Bulgaria, 172

Buhmann, M., 125

Bunsen, R., xxi

Butterfly diagram (see Maunder, E.W.)

C

Calcium, 78, 115

Canada, 170, 172, 180

Cannon, A.J., xvii, xxi

Carbon

- As fuel (see fossil fuel)

- Atom, 7 (structure) 78

- Dioxide, , 7, 22-24, 68-69 (use by plants for energy) 159 (re-absorption by cement) 159 (see also Biosphere I and II)

- Isotope Carbon-14 [C14]) 22-23, 25-29, 36, 38, 42, 44, 46, 81, 110-113, 174, 179

Cardinal (as permanent northern U.S. migrant) 48

Carnot, N.S., xxiii

Carrington, R.C., 14-15, 75, 99, 100

- Carrington's Law (see also Spörer's Law [Sporer's Law]) 88, 155

Centers of activity, 81-84, 86-88, 92-93, 95, 98, 103, 108, 115-116, 118, 120

Central Africa, 42-43

Central Asia, 43

Central Russia, 42

Cepheus, xiv

Cepheid variables, xiv

CERN, 6, 64, 76-77, 157-158

Ch'ing (see Qing Dynasty)

Chadwick, J., 5

Chandrasekhar, S., xxiv-xxv

Chapman, S., xxvi, 12, 16, 61-62

Charlemagne, 50

Charged particles (see particle coupling)

Chéseaux (Cheseaux) J-P. Loys de, xvii

Chaotic transitions (or events), 127, 131, 141

Chile, 44, 171

China, 53, 55, 104, 169-170

Chinese Dark Ages, 51

Chromosphere, 78, 86, 93, 96, 102, 115-116 (see also Layers of the sun)

Civilization (collapses of) 43, 50, 54-55, 178

Clausius, R., xxiii

Clerke, A., 12, 18-19, 26-27, 69-70, 112

Climate cyclicity, 112 (see also Hallstatt Cycle)

Climate Optimum (see Holocene Maximum)

Cliverd, M.A., 31, 39, 149

CLOUD (experiment) 76-77, 158

Cloud cover, 64, 69, 77, 157, 159, 162-163

Co-rotational (aspect of solar motion) 20, 96

Coal, xvi-xviii, xxii-xxiii, 3

Cold War, xxvi, 25-26

Colony Collapse Disorder (CCD) (see also Bees)
    178

Columbia University, 78, 159

Conflict (in context of possible climate
    influence) 31, 49-50, 55-58

Convection (Earth atmospheric) 65, 124, 158,
    163, 169 (solar) 84, 90, 118, 135

Core spin (solar) 128, 132, 141 (see also
    Gleissberg Cycle, inertial motion)

Coriolis Effect, 159

Corona, 72, 74, 85-86, 91, 93-94, 102, 150

Coronal Mass Ejections (CMEs) 61, 67-71, 81,
    84-86, 96, 98-100, 155 (halo type) 85

Coronal holes, 67-69, 94, 152

Corpuscular rays (see plasma)

Cosmic flux (see TSI)

Cosmic rays, xxvii, 5-6, 23, 27, 62, 64-65,
    67-69, 71, 74-76, 95, 100, 102, 110-11,
    151, 157-158, 161

Cosmogenic, 5, 39, 110

Cro Magnon, 27, 34

Crop failure, 45

Curie, M. xxii, 4
    • As unit (Tc) (Curie Temperature) 102

Cyclic (one of two quasi harmonics. See also
    Schwabe, Hale, Gleissberg, De Vries,
    Hallstatt Cycles) 112

Cyclonic activity, 57, 170 (see also ENSO)

Czech Republic, 172

D

D layer (ionosphere) 71, 74, 77, 153

D' Aleo, J., 161

DC (direct current) 2

DM (dipolar maximum) 135

DMmax (dipolar field strength) 135, 143

Da Vinci, L, 18-19

Dalton, J., 4

Dalton Minimum, 42, 52, 110, 137, 141, 149,
    150, 168, 175

Darwin, C., xxv, 1-4, 7-8

Darkening component (of measured TSI) 108

Dansgaard–Oeschger events (see Bond, G.)

Deflective mechanism (see also reflective
    mechanism) 61

Delta T (ΔT, averaged temperature change) 79,
    81, 109

de Jager, C., 31, 39, 80, 88, 96, 102, 108, 112,
    116, 120, 125, 132, 135, 141, 145, 150, 156,
    162, 168, 176, 177, 181

de Laat, A. T. J., 78

De Magnete, 101

de Vries, H., 23-26
    • De Vries Cycle (see also Suess Cycle),
      127-129, 131-133, 152, 179

Dendrochronology, 16

Descartes, R., xvi, xxviii, 96

Dessler, A.E., 161

Differential (motion) 68, 90, 92, 105, 113-114,
    127-128, 140, 163-164

Dipole (see also magnetic dipole, dipolar field,
    Hertz, H.) 85, 120

Douglass, A.E., 16-19, 23-24, 26, 51

Dove, Mourning (as permanent northern
    migrant) 48

Drought, 31, 43-45, 50-51, 169-171, 173, 175-
    176, 178, 180 (tables) 55

Duhau, S., 31, 80, 88, 102, 108, 112, 116, 120,
    125, 128, 135, 141, 145, 150, 168, 177

Dynamism, x, xxv, 4

Dynamo (see solar dynamo)

Dynamo theory (see solar dynamo)

E

Earth radii, 72, 91, 94, 96

Earthquakes, 29

Eccentricity, 78, 105-107, 124

Eddington, A., xxiv-xxv

Eddy, J., 21-23, 26-27, 28, 46, 146

Egypt (Egyptian) 50

Einstein, A., xxiv, 124, 178

El NinÕ Southern Oscillation (ENSO) 55-59, 154, 159-161, 163, 171

Electromagnetic spectrum (see UV, EUV, microwave, IR, gamma, Á and B radiation, X-ray)

Electrosphere, 73-75, 77, 153 (see also stratosphere and troposphere)

Energetic Emission Delay (or Effect) (EED) 86, 99, 118

Energy transport, 104, 106, 111, 174

Ephemeral Active Regions (EARs) 155

Episode (see grand episodes and grand phases)

Episodic (solar magnetic short term fluctuations: one of two quasi-harmonics) 112

Escape velocity, 61

Etruscan, 50

Eudoxus, xvi

Europe, 43, 47, , 104, 172, 177

eV (electron volts) 95, 98

Exosphere (diagram of) 71

Extreme Ultraviolet (EUV) (see also electromagnetic spectrum) 85, 92-93, 102-103, 152, 154, 158, 162-163

Extreme Ultraviolet Variability Experiment (EVE) (a subsystem of SDO) 72, 85, 103, 162

F

Φ record (long and short term solar fluctuation model) 112-113

F layer (ionosphere) 71

Facula/Faculae (see also centers of activity) 81, 86-87, 108, 115, 152

Faeroes, 27

Fairbridge, R., 133, 171

Faraday, M., xxiii-xxvi, 90

Ferraro, V.C., 62

Field(s) (see magnetic fields)

Finch, house (or Mexican, as permanent northern migrant) 48

Flocculi (Hale's solar definition) 89 (see also molecular Hydrogen)

Floods, 43, 45

Forbush, E.H., 48

Fossil fuel, 158, 172, 180

Foukal, P.V., (lack of full solar activity theory) 60 (see also Landscheidt, T.)

Fourier, J., (also Fourier) xix, xxvi, 31, 124-125

Fourth state of matter (as a fourth putative change of matter's state. See plasma)

Fox squirrel, Eastern (as possible Holocene Maximum northward migrant) 36

Frölich, C., (Frolich) 108, 146

Fukuda Scale (see also magnitude) 171

G

G class (of stars) xiv, 152

Galactic Cosmic Rays (GCRs) 75, 77-78

Galileo, xv, xvi-xviii, xx, xxviii, 128, 134

Galactic Ray Bursters (GRBs) 153

Galactic supernovae (see supernovae; GCRs)

Galerkan methods, 126

Gamma rays, radiation (see also electromagnetic spectrum) 67, 74, 93-94, 101, 152

Gauss, C.F., xx, 13-14
- As unit (Gauss) 86, 96, 122, 152

Geological Time Scale, 8

Geomagnetism, iii (storms) xxiv, 10, 13, 86, 99 (Earth's geomagnetic field) 100, 111, 113, 129

German Academy of Science, xxvi

Germany, 12, 14

Ghana, 56

Gilbert, W., xvi, 101.153

Glaciers, glaciation, 33, 45-46, 65, 73

Gleissberg Cycle, 60, 112, 127-133, 137, 139-145, 152, 154, 168

Global Climate Models (GCMs) 157, 160

Gnevyshev Gap, 99

God, 8, 40, 179

Godwin, H., 23

Göttingen (Gottingen) University, xx

Gradients, 65-66, 104, 163, 174 (see also solar and orbital forcing)

Grand episodes (extended minima) 39, 47, 50, 127, 131-132, 135, 141, 150, 169

Grand phase(s) (of sun, usually extended maxima) 10-12, 69, 70, 116, 131-134, 136-138, 141, 143-144, 149

Gravity, gravitational, 106, 226, 133, 163

Greece, 56, 273

Greeks (classical) xviii

Greenland, 27, 42
- Greenland Low Pressure System, 65-66, 159 (see also gradients)

Greenhouse Effect, 78, 124, 157-158

Greenhouse Gases (GHGs) 68, 78, 141, 152, 157-158

Greenwich, xxiii, xxvi, 14

H

Hadley Cell, 43-44, 65-66, 154, 163, 171

Hale, G. E., 89-90
- Hale Cycle, 10, 87-88, 112, 127-133, 144-145, 152, 154, 156

Halo CMEs (see CMEs)

Hallstatt Cycle, 31, 28-39, 112, 128, 142, 149-150, 156, 165, 169, 175, 177

Harvard Smithsonian Center for Astrophysics, 22

Helium, xvi, xvii, 98

Heliopause, 63-64, 66

Helioseismology, 83

Heliosphere, 61-64, 68-69, 72, 74-75, 91, 94-98, 118, 151

Helmholtz, H. von, xxi, 3

Hematite (stains) 33, 35

Henry's Law (or coefficient) 35

Hercules, 61

Herschel, C., xviii

Herschel, J., xx

Herschel, W., xviii, xix, xx, 14

Hertz, H., 88-89, 120, 126

Hertzsprung, E., xiv
- Hertzsprung-Russell diagram, xiv

High Altitude Observatory (HAO) 22

Higgs Field (Boson) 123

Hill, F., 123, 155-156

Hobarton (watch station) 13

Holocene, Holocene Maximum, 32-38, 41-45, 103, 106-107, 110-111, 155

Houghton, J., 159

Howell, W.N., 31, 41, 45, 50-51, 149 (see also pseudo-decadal averaging)

HR diagram (see Hertzsprung-Russell diagram)

Hsiang, S., 31, 56-58, 175

Hubble (telescope) xvi

Humboldt, A. von, 13

Hurricanes (see cyclonic activity)

Hydrogen, xiv, 3, 72, 75, 87, 98

- Atomic, 89
- Molecular, 89 (see also flocculi)

Hydrodynamic, 61-62, 64, 67-69, 72
Hydrology, hydrological, 41, 45, 163
Hydromagnetic, xxvi, 96, 117
Hydrostatic, 61-62, 64, 72

I

Icebergs, 33
Ice Age, 25, 31-35, 40 , 45
Ice cores (archive or reservoir) 100, 110-111
India, 42, 33, 50-51
Indian monsoon (cycle) xxiv, 57
Indian Ocean, 171
Inertial motion (solar) 86, 90, 95, 120, 127, 130, 132-133, 137-140, 154
Infra Red (radiation) (IR) (see also electromagnetic spectrum) 24, 65-70, 78, 84, 92-93, 152, 157-158, 179
Inner rotation (solar core) 20, 84 (see also inertial motion)
Insecticides, 178
Insolation (see solar insolation)
Interstadials, 33
Ionosphere, 71, 6, 23, 72, 74-75, 85, 97, 103, 116, 152-153
Ions, ionization, 5, 74-75, 94-95, 119, 158
IPCC, 159-161
Irradiance (see TSI)
Isotopes , 1-8 (explanation of) 16-19, 22-26
Isotropic (see linearity)
International Geophysical Year (IGY) 176

J

Jastrow, R., 7
Job, Book of, 40, 180
Jupiter, 63, 106, 133

K

Kant, I., xviii, xix
Kapteyn, J., xix, 12
Kataja, E., 129, 136
Kelvin , Lord (William Thomson, Baron of Largs) xviii, xix, xx-xxiv, 2-5, 7-8, 14, 21, 80, 117, 123-124
- As unit (Kelvins [K]) 72-73, 85-86, 96, 164
Kepler, J. xix, xvi, xxviii, 106, 151
Kew (magnetic observatory) 14, 99
Kilopascal, 66
Kirchhoff, G., xxi, 6
Kirkby, J., 6, 75-76, 157-158 (CLOUD experiment) 75-76
Krause, F., xxv, xxvi
Kuiper Belt, 24, 73
Kyr (kiloyear) 34-36

L

La Niña, 56-57, 171
Lake Champlain, 170
Lamont-Doherty Earth Observatory (Columbia University) 78
Lamoka, 36-37, 43
Landscheidt, T., 57-59, 80
Laos, 56
Latvia, 56, 172
Laurentian, 10, 34-35, 37
Laws of sunspot motion (see also Carrington's and Sporer's Law) 14-15, 85, 87-88
Layers of the Sun, 84, 95, 203, 115, 127, 151
Lead-206, 7
Leighton, R.B. (see also Babcock-Leighton) xxv
Lean, J., 93, 108, 146
Leavitt, H. S., xiv-xv
Lefroy, J., 21

Libby, W., 22-24

Light Year, xiii

Linearity xxiii, 3, 14 (as opposed to nonlinearity) 3, 77-78 (solar) 113-116 (in relation to TSI)

Lithosphere, 97, 104, 153-154

Little Ice Age (LIA) 34-35, 40, 45-48, 51, 163

Living And Working In Space (LAWIS) xxvii

Livingston, W., 122

Local (climate effects, extended to wider matrix) (see also regional climate effects) 40-49

Lockwood, M., 39, 134-139, 143

• Lockwood data (time series, etc. See also Standard data) 134-137, 143

Long wave radiation (see IR)

Longer term cycles ("solar") (see De Vries, Hallstatt Cycles)

Lord Kelvin (see Kelvin)

Lorenz force, 120-122, 141

Los Angeles, 172

Lower Gleissberg, 128-129, 131

Lowell Observatory, 16

Luminosity (see also TSI; spectral class) xiii (tie-in to spectral class) xiv (solar) 106

Lyman, T., 5

M

Macula. maculae (see sunspots)

Madden-Julian Oscillation, 161

Maddock, A., 23

Magnetic fields (bird olfactory glands sensitive to) 48-49 (general) 61, 64, 75, 84-85, 87, 92, 117

Magnesium, 115

Magnetohydrodynamics (MHD) xxvi, 62, 117-118, 151

Magnetopause, 63-64

Magnetosheath, 70-72, 75, 97, 100, 153

Magnetosphere, 6, 63, 26, 27, 61-626, 67, 69, 72-73, 77, 97, 99-100

Magnetotail, 64, 97

Magnitude (as force vectors: see Richter Scale, Fukuda Scale, Saffir-Simpson Scale, nT, etc.)

Mammoth, Wooly, 33

Mars, 63-64, 78

• Mars Society, xxviii

• Mars Academy, xxviii

Manhattan Project, 22

Matthes, F.W., 45

Maunder, E. W., xviii-xxvi, 4, 14, 20, 22, 57, 70, 88, 110, 112, 115, 117, 120, 121, 126, 137-139

Maunder Minimum, 10, 22, 26, 27, 29, 39, 44, 46-47, 51, 54, 140-141, 143-144, 147-150, 154, 156, 169, 170-174

Maunder, A.D.R., xxvi, 4, 12, 18

Massachusetts, 22, 48, 178

Maurellis, A.N, 78

Maurya Dynasty, 51

Maury, A.C., xiv, xvii

Maxima (solar cyclic high or active phases) (see Modern Maximum, Medieval Maximum, etc.)

Maxwell, J.C, , xx, xxi, xxiii-xxvi, 117

Mayer, J.R. von, xix, xx, xxiii, 5

MeV (multiples of electron volts) 98, 111-113, 122

Medieval Maximum (or Warm Period), 10, 26, 50-51, 57, 29, 37, 155

Mediterranean, 50, 53, 174

Mendeleev, D., XXI-XXII, 4

Meridional oscillation (solar) 81, 118

Mesoamerica, 50

Mesopotamia, Mesopotamian, 50

Mesosphere, 71, 72-74, 85, 153, 161

Messier, C., XVIII

Methane, 36 (see also GHGs)

Methodist, Methodism, 15

Meteoric (or meteorite) theory of solar power, xx, xxiii

Microclimate (see also regional and local climate effects), 36, 66-70, 175, 181

Microwave (radiation) 101, 152 (see also electromagnetic spectrum)

Middle atmosphere (see electrosphere)

Milankovitch Cycle, 78.106

Miller, S., 157

Milky Way Galaxy, xxviii

Mississippian, 50

Modern Maximum, ix, 26, 141-143, 148, 155

Mongol Empire, 51

Mongolian Low Pressure System, 159

Monthly Notices of the Royal Astronomical Society (MNRAS) 15, 57

Moon (Earth's), xvi, xxvi, xxvii, 73, 163

Muslims, xvi

N

nT (Southward Interplanetary Magnetic Field), 97, 134

Nagovitsyn, Yu. A., 129, 137, 148, 150

nm (nanometer) 87, 103

National Aeronautics and Space Administration (NASA) xxviii, 22, 48, 75, 103, 120-121

National Oceanic and Atmospheric Administration (NOAA) 134, 171, 173

Nature, 56

Navier, C-L, xix, xxvi

Navier-Stokes, 117-119, 124

Neanderthal, 27, 33-34

Netherlands, xix, 24 ("850 B.C. Event") 41-42, 46, 50, 51

Nesme-Ribes, E., 121

New Delhi, 179

New Jersey ("Superstorm Sandy") 178

New York State, 35-37

New Zealand, 172

Newton, I. xvi, 3, 80, 82, 132

Nitrates (as result of geomagnetic storms and filling Earth reservoirs or archives) 100

Nitric acid (adding to combustible nature of lower atmosphere) 74

Nitrogen, 105, 110

- As isotope Nitrogen-14 (14N) 7

Noctilucent clouds, 73

Non isotropic (see nonlinear, nonlinearity)

Nonlinear, nonlinearity, xxiv, 77, 103 (in relation to TSI) 113-116

Normal (or regular) phase, 10-11, 64-66, 70, 121-122, 127, 130-131, 134, `36, 138-141, 145, 154, 163-164, 177

North America, 51, 53

Northern Hemisphere, 10, 19, 27, 34, 36-37, 42, 46, 49-55, 64, 68-69, 78-79, 87-88, 99, 169-170, 172-173, 177

North, J., xxi

Nova Scotia, 27

Nuclear power, 6, 180

O

Obliquity, 105-107

Ohio Valley (see Mesoamerica)

Oklahoma, 170-171

Omega Effect, 92-94

Oort Minimum, 27, 39, 50-52, 54, 149

Orbital forcing, 96, 105-107, 164, 171 (see also eccentricity, obliquity, precession)

Orbital tilt (see solar forcing)

Ornithology, 47

Orography, orographical (wind, pressure in topology as climate influencer) 65, 105, 174

Oscillation (solar, of tachocline) 119-122 (in harmonics functions) 125, 132, 139

- Regular oscillations, bi-decadal semi-secular, 140-141, 144, 147-148, 154

- Earth oscillations: see ENSO and QBO)

Outer convection layer (solar) 90

Oxygen, 23

- As isotopes Oxygen-16 (O16) and Oxygen-18 (O18) 32-35

- Comparisons with C14, 23

P

Pacific Ocean, 57, 180 (see also ENSO)

Pakistan, 171, 180

Palynology, palynological, 37, 42-43

Paris (France) 179

Parker, E.N., xxvi, 94

Parker Spiral, 6, 64

Particles (see cosmic rays, particle coupling)

Particle accelerator, 97, 158

Particle coupling (explanation of) 77, 116 (possible climate influence from) 109

Pascal, B., 174

- As unit (Pascals [Pa] and Kilopascals [Kpa]) (see gradients)

Paschen, F., 5

Penn, M., 172

Pennsylvania, 36

Periodic table (of elements) viii; xxi-xxii, 4, 6

Persian Seleucid, 51

Peru, 171

Philippines, 34

Phase diagrams, 144, 146, 149

Phoenician, 50

Photosphere, 84, 86, 93, 113, 115

Photovoltaics, panels (see solar power)

Pions (see also CLOUD experiment; CERN; plasma; particle coupling) 76, 58

Plages, 81, 84, 86, 152 (see also centers of activity)

Plagues, 44-45, 50

Planets (Solar System) xxvii, , 63, 70, 97, 105-106, 153, 164 (as not adding force to Sun) 132-133

Poland, 172

Polar Facula Regions (PFRs) 88, 118, 120

Polarity, 87-89, 126, 175

Poloid, poloidal (see also proxies for minina; aamin; aa index), 90-92, 118, 130-131, 136-138, 164

Plages (see also centers of activity)

Plasma, 61, 67-72, 80, 83-85, 94-97, 102, 108, 120, 123-124, 152

Pleistocene, 45, 107

Portugal, 52

Precession, 105-107

Pressure (see gradients)

Prominences, 93, 95-96, 102, 152 (see also centers of activity; chromosphere)

Proton events (see SEPs)

Proxy evidence (see isotopes; [solar] Rmax; aamin; sun-like stars)

Pseudo-decadal averaging, 32, 36, 44-45, 49, 51

Ptolemy, xvi

Pulsating variable (star) (see also variable stars)

- Our Sun being a possible type of, xv

Q

Qing Dynasty, , 51

Quantum (mechanics) xxi, 5

Qualitative (appreciative conceptual measure) 30

Quantification (quantitative [numerical] not specifically statistical, measure) 122, 169

Quasi Biennial Oscillation (QBO) 163

Quasi-harmonics (or, cyclic and episodic phases) 112, 130-132, 137, 139, 141, 143

Quebec, 180

Quiet regions, 94

R

R0 (solar radii) 94

R (as sunspot number) 114 (see also Zurich-Wolf number)

R (for Regular solar phase, or, oscillation: see Normal)

Rmax 129, 131, 136-139, 143 (see also toroid)

Radial basis functions of compact support, 114, 126

Radio emission (with UV from chromosphere) 115

Radioactive decay, 180

Radioactivity 4 (see also Radium; isotopes; radionucleids; particle coupling; cosmogenic; anthropogenic)

Radiocarbon (dating) (see also Carbon-14)

Radionucleids (see isotopes; radioactivity; particle coupling)

Radium, 5

Reconnection, 95

Reflective mechanism, 61

Regional climate effects (see also local climate effects) 33, 40, 42, 45, 48, 66, 79, 99, 152, 158, 169

Regular (R) (phase or solar oscillation versus grand phase: see Normal)

Relativistic particles, 67, 96-98, 118 (see also SEPs)

Reservoirs (see trees, ice cores for various isotopes)

Richter Scale (see also magnitude (force vectors) 97, 171

Ritchie, W., 36-37 (see also archaeological evidence for deep-time global climate change; C14)

Rhine River, 51

Roman Empire, 50-51

Romania, 172

Rome, xvi, 172

Royal Astronomical Society (RAS) 4, 14-15, 57, 99

Royal Observatory (see Greenwich)

Royal Society, xx, 8, 23

Rushid, Ibn al, xvi

Russell, H.N., xiv

Russia, 42, 58, 169, 171, 173, 178

Rutherford, E., xxiii, 4-5

S

Sabine, E., xx, xxi

Sadi-Carnot, N., xxiii

Saffir Simpson Scale (see also magnitude) 171

Saturn, xvi, 106, 133

Scheiner, C., 134

Schwabe, S.H., 12-14

- Schwabe Cycle, 10, 16-17, 19, 24, 26, 29, 38, 58, 87-88, 108, 110, 112-113, 127-130, 144-145, 147-148, 152, 155-156, 162

Schove, J.D., 38-39, 150

Scythian, 43

Seattle, 172

Semi-secular oscillations, 131, 139, 145, 148

Severinghaus, J.P., 78, 159

Short term solar fluctuations, 112-114 (see also quasi harmonics, cyclic and episodic)

Siberia, 37, 43

Siderenkov, N.S., 163

Shock waves (see bow shock, nT)

Short wave radiation (see visible light)

Slovakia, 172

Small Finger Cycles (SFCs) 58-59

Smithsonian Institution, 22, 26

Smithsonian Center for Astrophysics, 22

Snowfall, 172 (see also respective countries recently heavily effected [Slovakia, Czech Republic, etc.])

Sodium chloride (in tests for C14) 23

Sodium hydroxide (in tests for C14) 23

Solanki, S.K., 38

Solar and Heliospheric Observatory (SOHO), xxviii, 62, 64, 118

Solar constant, xx, 81, 102, 107, -108, 118 (see also TSI)

Solar Dynamics Observatory (SDO) xxviii, 85, 93, 95, 103, 108, 113-114, 163

Solar cycles (see individual solar cycles as listed)

Solar dynamo (theory), xxv-xxvi, 62, 86, 102, 108, 113, 116-117, 124, 128, 133, 135, 141

Solar Energetic (or Energized) Particles (SEPs) 67, 69, 72, 75, 80, 84, 86, 91, 94-95, 98100, 114, 116, 123, 124

Solar flares, 67-70, 81, 86, 93, 95-96, 98-99, 103 (classes of) 84-85 (see also centers of activity; CMEs)

Solar forcing, 40, 42, 107, 110, 163 (see also orbital forcing; solar luminosity, orbital tilt)

Solar power (solar energy) 181

Solar wind, 27, 61, 63, 67-69, 72-73, 85, 94, 96, 99, 152, 161

South Africa, 173

Southern Oscillation Extreme Patterns (SOI) 58

Spain, 52

Spectral class, xvi-xvii, xxi (see also luminosity)

Spenser, R.W., 161, 171

Spectral Solar Irradiance (SSI) 86, 102-103, 109-110, 113-114, 123, 162 (see also TSI)

Spectroheliograph, 89 (see also Hale, G.E.)

Spinoza, B., 8

Spörer, G. (Sporer) 12-15

Sporer's Law, 15, 88-89, 155 (see also Carrington's Law)

Sporer Minimum, 17, 26-27, 39, 46-47, 51, 54, 149

Spradling, M., 170

Standard data (contra Lockwood data) 39, 134-136, 143

Steenbeck, M., xxv-xxvi

Steinhilber, F., 31, 38, 39, 149

Stellar activity (see centers of activity)

Stasis (as opposite of dynamism) 64, 92

Statistics (see also wiggle matching, Delta T)

Stratosphere, 71 (diagram) 66, 72, 74, 85, 102-103, 105, 116, 152-154, 161-164

Stockholm International Peace Research Institute (SIPRI) 25

Sudan, 56

Suess Cycle (see De Vries [or de Vries] Cycle)

Sui Dynasty, 51

Sulphuric acid, 75 (see also CLOUD experiment)

Sun-like stars, xiv

Sunspot cycles (see laws of sunspot motion, sunspot groups)

Sunspot groups , 81 (as calculated with TSI) 164-168 (charts of last 11 solar cycles) 15, 83, 88, 91, 107, 121

Sunspots, xvii-xviii, xx, 14, 28, 57, 68-69, 81, 83-84, 86-89, 95, 107-108, 115, 118, 120-122, 130-131, 141, 152 155-156, 164-165 (see also centers of activity; bearers)

Supernova, supernovae (exploding stars) xv, 6, 63-64, 68, 74

Svalgaard, L, 134

Svensmark, H., 64, 161

Sweden, 56

Switzerland, 172

T

Tachocline, 84-87, 90-94, 100, 118-120, 127-128, 130-133, 135, 137, 139-141, 154

Tang Dynasty, 51

Tesla, N., 2

Texas, 48, 170-171

Titmouse, tufted (as permanent migrant northward) 48

Thames River, 47

The Economist, 155

Thermoclines, 104

Thermodynamics, xx-xxii, xxvi, 35, 123

Thermohaline circulation, 36, 209

Thermosphere 71 (diagram of) 72-74, 153, 161

Thompson, M., 7

Thomson, W. (see Kelvin)

Three laws of planetary motion, xix

Thunderstorms, 74, 100

Toroid, toroidal (see also proxies for maxima; Rmax) 81, 90-92, 118-119, 129-131, 133-134, 136-138, 164

Tornados (see cyclonic activity)

Toronto (watch station), 13, 21

Torsional oscillation (internal twisting motion of sun) (in solar dynamo) 120, 127, 139-141

Total Solar Irradiance (TSI) xx, 28, 32, 68, 69, 79, 86, 94, 102-103, 107, 123-124

- As variable with sunspot motion, 81, 93, influence on upper atmospheric chemistry 109-112

Transition point , 128-129, 134, `36-137, 139-140, 144-145, 148, 150, 168

Transition Region Coronal Explorer (TRACE) xxviii

Trees (as isotope reservoirs or archives or both) 7, 16-18, 23, 27, 42, 51, 73, 97

Trinidad, 56

Trenberth, K, 160-161

Troposphere, 71 (diagram) 64, 72-76, 79, 84, 93, 100, 105, 115-116, 153, 158, 161

Tunisia, 56

Turkey, 173

Type 1A supernova (see supernova, supernovae)

TÜbingen (Tubingen) University, xx

U

Ultra Violet (radiation) (UV) 67-68, 70, 72, 74, 78, 84-85, 92-93, 95, 102, 115, 152-153, 158 (see also EUV)

Ukraine, 172

Ulysses, xxvii

Unipolar Regions (URs) 88, 118

United Nations, 26 (see also IPCC)

Uranium-235, 7-8, 23

Usoskin. J.L., 38-39, 113, 150

U.S. Drought Monitor, 170

U.S. National Solar Observatory, 155

U.S. Naval Academy, 21

University of Colorado, 21

University of Arizona, 16

University of Gröningen (Groningen University) 23

Upper Gleissberg, 128-129, 131, 139

V

V/m fields, 73

Van Allen Belts, 61

Van Geel, B., 41-46, 50, 103, 116, 171

Variable stars (see also pulsating variable) xiii-xv, xxvii, 127, 152-153, 164, 181

Venturi tube, 144

Venus , 6, 63-65, 105, 124, 158-159

Vermont, 16, 49, 174

Vikings, 27, 37, 42, 70

Vinland, 70

Visible light, 3 (low energy) 45, 158 (see also electromagnetic spectrum)

Z

Zhou Interregnum, , 51

Zurich-Wolf Number (sunspot number; see also R for sunspots) 13

W

W/m2 (in solar strength in pseudo-decadal averaging) 41, 49-55

- In revised data on declining Schwabe Cycles since the 1990s, 145

Washington, D.C., 22

Water vapor , 35, 50, 78, 141, 157, 175 (see also GHGs)

Wavelength (see electromagnetic spectrum),

Wiggle matching, 42 (see also Geel, van B.)

Wilson, E.O., xxvii

Wolf, R., 13

- Wolf Minimum, 17, 26-27, 39, 52, 54, 129, 149

Woodland, 50

World War I, xxv, 12, 14

World War II, xxv-xxvi, 22, 25, 177

Wright, T., xviii, xix

X

X-ray, 67, 74 (high energy) 67-74 (general) 70, 74, 92, 94, 101-103 (see also electromagnetic spectrum)

Y

Ybp (Years Before Present) 34, 104

Yohkoh, xxviii

Younger Dryas, 35

Yuan Dynasty, 51